# RESEARCH AND PERSPECTIVES IN NEUROSCIENCES

Fondation Ipsen

*Editor*

*Yves Christen*, Fondation Ipsen, Paris (France)

*Editorial Board*

# Springer

Berlin
Heidelberg
New York
Hong Kong
London
Milan
Paris
Tokyo

J. Mallet   Y. Christen (Eds.)

# Neurosciences at the Postgenomic Era

With 13 Figures and 8 Tables

 Springer

*Mallet, Jacques,* Ph. D.
Laboratoire de Génétique Moléculaire de la Neurotransmission et
des Processus Neurodégénératifs
Centre National de la Recherche Scientifique
UMR, Bât CERVI
Hôpital Pitié Salpétrière
75013 Paris
France
e-mail: mallet@infobiogen.fr

*Christen, Yves,* Ph. D.
Fondation IPSEN
Pour la Recherche Thérapeutique
24, rue Erlanger
75781 Paris Cedex 16
France
e-mail: yves.christen@beaufour-ipsen.com

IISBN 3-540-00194-8  Springer-Verlag Berlin Heidelberg New York

Library of Congress-Cataloging-in-Publication-Data
Neurosciences at the psotgenomic era / J. Mallet, Y. Christen (eds.). p. cm. – (Research and perspectives in neurosciences) Includes bibliographical references and index.
  ISBN 3-540-00194-8 (hardcover: alk. paper)
1. Neurogenetics. 2. Behavior genetics. 3. Nervous system-Diseases-Genetic aspects. 4. Nervous system-Diseases-Gene therapy. I. Mallet, J. (Jaques) II. Christen, Yves. III. Series.

Springer-Verlag a member of BertelsmannSpringer
Science + Business Media GmbH

http://www.springer.de

© Springer-Verlag Berlin  Heidelberg 2003
Printed in Germany

The use of general descriptive names, registered names, trademarks, etc. in this publication does not imply, even in the absence of a specific statement, that such names are exempt from the relevant protective laws and regulations and therefore free for general use.

Product liability: The publishers cannot guarantee the accuracy of any information about dosage and applicat12ion contained in this book. In every individual case the user must check such information by consulting the relevant literature.

Production: PRO EDIT GmbH, 69126 Heidelberg, Germany
Cover design: design & production, 69121 Heidelberg, Germany
Typesetting: AM-productions GmbH, 69168 Wiesloch, Germany
Printed on acid-free paper        27/3150Re – 5 4 3 2 1 0

# Contents

# Contributors

*Barkats, M.*
Laboratoire de Génétique Moléculaire de la Neurotransmission
et des Processus Neurodégénératifs,
Centre National de la Recherche Scientifique UMR 7091,
Bât CERVI, Hôpital Pitié Salpétrière,
75013 Paris, France

*Bemelmans, A.*
Laboratoire de Génétique Moléculaire de la Neurotransmission
et des Processus Neurodégénératifs,
Centre National de la Recherche Scientifique UMR 7091,
Bât CERVI, Hôpital Pitié Salpétrière,
75013 Paris, France

*Botas, J.*
Department of Molecular and Human Genetics,
Baylor College of Medicine, One Baylor Plaza, Houston, TX 77030, USA

*Brun, S.*
Laboratoire de Génétique Moléculaire de la Neurotransmission
et des Processus Neurodégénératifs,
 Centre National de la Recherche Scientifique UMR 7091,
Bât CERVI, Hôpital Pitié Salpétrière,
75013 Paris, France

*Caron, M.G.*
Howard Hughes Medical Institute and Department of Cell Biology,
Duke University Medical Center, Durham, NC 27710, USA

*Chelly, J.*
Laboratoire de Génétique et Physiopathologie des Retards Mentaux, ICGM,
CHU Cochin, 75014 Paris, France

*Fisher, S.E.*
Wellcome Trust Centre for Human Genetics, Oxford University,
Roosevelt Drive, Oxford, OX3 7BN, UK

*Gainetdinov, R.R.*
Howard Hughes Medical Institute and Department of Cell Biology,
Duke University Medical Center, Durham, NC 27710, USA

*Giustetto, M.*
Howard Hughes Medical Institute, 1051 Riverside Drive, New York,
NY 10032, USA

*Grant, S.G.N.*
Department of Neuroscience, University of Edinburgh, 1 George Square,
Edinburgh EH8 9JZ, UK

*Guan, Z.*
Center for Neurobiology and Behaviour,
College of Physicians and Surgeons, Columbia University,
New York State Psychiatric Institute, 1051 Riverside Drive,New York, NY 10032,
USA

*Kandel, E.R.*
Howard Hughes Medical Institute, 1051 Riverside Drive,
and Center for Neurobiology and Behaviour,
College of Physicians and Surgeons, Columbia University,
New York State Psychiatric Institute, 1051 Riverside Drive,
New York, NY 10032, USA

*Levitt, P.*
Departments of Neurobiology and PittArray,
University of Pittsburgh School of Medicine, W1655 BST, Pittsburgh,
PA 15261, USA

*Lewis, D.A.*
Departments of Psychiatry and Neuroscience,
University of Pittsburgh School of Medicine, W1655 BST, Pittsburgh,
PA 15261, USA

*Lomvardas, S.*
Department of Biochemistry and Molecular Biophysics,
College of Physicians and Surgeons, Columbia University,
630 West 168th Street, New York, NY 10032, USA

*Mallet, J.*
Laboratoire de Génétique Moléculaire de la Neurotransmission
et des Processus Neurodégénératifs,
Centre National de la Recherche Scientifique UMR 7091,
Bât CERVI, Hôpital Pitié Salpétrière, 83 Bd de l'Hôpital, 75013 Paris, France

*Mandel, J.-L.*
Institut de Génétique et de Biologie Moléculaire et Cellulaire,
CNRS/INSERM/Université Louis Pasteur, Illkirch, CHU Strasbourg, France

*Middleton, F.A.*
Department of Neurobiology, University of Pittsburgh School of Medicine,
W1655 BST, Pittsburgh, PA 15261, USA

*Mirnics, K.*
Departments of Psychiatry, Neurobiology, and PittArray,
University of Pittsburgh School of Medicine, W1655 BST, Pittsburgh,
PA 15261, USA

*O'Dell, T.J.*
Department of Physiology and Interdepartmental Ph.D. Program
for Neuroscience, UCLA School of Medicine, Los Angeles, CA, 90095, USA

*Pierri, J.N.*
Department of Psychiatry, University of Pittsburgh School of Medicine,
W1655 BST, Pittsburgh, PA 15261, USA

*Plomin, R.*
Social, Genetic and Developmental Psychiatry Research Centre,
Institute of Psychiatry, Kings College London, 111 Denmark Hill,
London SE5 8AF, UK

*Sarkis, C.*
Laboratoire de Génétique Moléculaire de la Neurotransmission
et des Processus Neurodégénératifs,
Centre National de la Recherche Scientifique UMR 7091,
Bât CERVI, Hôpital Pitié Salpétrière,
75013 Paris, France

*Schwartz, J.H.*
Center for Neurobiology and Behaviour,
College of Physicians and Surgeons, Columbia University,
New York State Psychiatric Institute, 1051 Riverside Drive,
New York, NY 10032, USA

*Thanos, D.*
Department of Biochemistry and Molecular Biophysics, College
of Physicians and Surgeons, Columbia University,
630 West 168th Street, New York, NY 10032, USA

*Verma, I.M.*
The Laboratory of Genetics, the Salk Institute,
10010 North Torrey Pines Road, La Jolla, CA 92037, USA

# Introduction: Neurogenomining

Y. Christen[1] and J. Mallet[2]

The Human Genome Project constitutes a major revolution in scientific research, and the publication in February 2001 of the first draft of the human genome (Venter et al. 2001; International Human Genome Sequencing Consortium 2001) was a symbolic event that the future will note. This scientific revolution has consequences for every domain of biomedicine but perhaps especially for the most complex among them and thus, particularly, for the neurosciences. For this reason the Fondation Ipsen devoted one of its Colloques Médecine et Recherche to the mining of the human genome, prior to the end of this symbolic year. This encounter took place in Paris on 3 December 3, 2001.

Neurogenomics in all of its forms had expanded considerably even before the decoding of the human genome, but it will benefit greatly from the knowledge gained from this sequencing. The progress simultaneously concerns the development of new research tools, as well as knowledge of diseases and of neurophysiological and cognitive processes and the domain of treatment, – present and future. The Ipsen Foundation Colloquium made it possible to focus on several of these areas.

## The Decoding: a Beginning More than an End

The publication of the draft sequence of the human genome, simultaneously in *Nature* and *Science,* is only symbolically an achievement. Not only this draft remains incomplete, but also the data obtained have yet to be interpreted. The publication of the sequence does not even make it possible to individualize, or identify, the genes, to know where they begin and end, or even to count them precisely. At most, we now know that their number is much lower than many researchers had thought – on the order of 30,000 or 40,000 (Claverie 2001). Even once they have been individualized, these genes will require substantial study before their functions will have been identified, a work which will take years and will certainly provide numerous surprises.

Moreover, the Human Genome Project has decoded only a reference genome (in the same sense as we might speak of a reference or standard meter). The

[1] Fondation Ipsen, 24 rue Erlanger, 75016 Paris, France
[2] Génétique Moléculaire et Neurotransmission, Bâtiment Cervi, Hôpital Pitié-Salpêtrière, 83 Bd de l'Hôpital, 75013 Paris, France

Mallet/Christen
Neurosciencess at the Postgenomic Era
© Springer-Verlag Berlin Heidelberg 2003

interaction between genes and environment causes enormous variability among human beings. Each copy of the human genome is unique and differs in sequence from any other copy in the population by roughly 1 per 1,250 nucleotides (Sachidanandam et al. 2001). Genetic polymorphism is not only related to diseases, of course. Most individual differences between individuals have no pathological consequences and, in many cases, we do not know their physiological meaning. They nonetheless influence physical and psychological characteristics as well as predisposition to diseases which is why it is as important to know the differences between individuals as the similarities. A major effort is therefore underway to identify the single nucleotide polymorphisms (SNPs). SNPs are single-base differences in the DNA sequence observed between individuals in the population. A polymorphism is defined as the least common allele occurring in 1% or more of the population, whereas mutations are rare differences that occur in less (usually much less) than 1% of the population (Roses 2000a). Mutations are more often discovered in coding sequences of genes that cause inherited diseases. SNPs are present throughout the genome. The SNP Consortium (a consortium of pharmaceutical and bioinformation companies, five academic centers and a charitable trust) is currently trying to produce an ordered high-density SNP map of the human genome (Roses 2000a).

From this point of view, the sequencing constitutes a beginning much more than a completion. According to Hardy (2001) "the accomplishment of a draft of a consensus sequence marks the beginning of a long and challenging road both scientifically and politically. Biology, medicine, and society will never be the same again."

We now face a tremendous task: mining the human genome. This enormous enterprise involves all aspects of biomedicine, especially the neurosciences. Nestler (2001) used the term psychogenomics "to describe the process of applying the powerful tools of genomics and proteomics to achieve a better understanding of the biological substrates of normal behavior and of diseases of the brain that manifest themselves as behavioral abnormalities."

## A Revolution More Technological than Philosophical

We are now beginning to have a better understanding  of ourselves based on molecular structure of the human genome. It is changing our view of ourselves (Pääbo 2001). For now, however, the genomic-postgenomic revolution, often presented as conceptual, is probably mainly methodologic. Beyond the actual sequencing, numerous tools have made it possible to approach the functional implications of our genetic machinery and are moving us from the genomic to the postgenomic era. Differential display technologies and DNA chip array technologies are only some of the approaches to gene expression in the central nervous system that are entering a phase in which they have become almost routine. These techniques can also be applied to single cells. Thus it becomes possible to use DNA arrays to examine, for example, the molecular biology of the dendrite (Eberwine et al. 2001).

In a completely different domain, the development of transgenic animal models of various cerebral diseases now extends to many species, no longer just only to mice. There are now numerous strains of *Caenorhabditis elegans* and *Drosophila* that carry harmful human mutations and reproduce at least some aspects of human diseases. These strains are particularly important because, in many cases (Alzheimer's disease, Huntington's disease, Rett syndrome), we had no serious model of these diseases before the discovery of the genes involved in their onset and the creation of transgenic animals that express these genes. In other cases (Parkinson's disease, addiction, schizophrenia), pharmacological models already existed but the contribution of genetic data has enabled the development of new and better models. These models improve our knowledge of the pathologies of these diseases, especially the mechanisms by which an accumulation of misfolded proteins exert toxic effects on neurons in most neurodegenerative disorders (Gusella and MacDonald 2000; Wong et al. 2002; Zoghbi and Botas 2002). They also provide information about the potential role of various neurotransmitter systems in symptoms (Caron and Gainetdinov, this volume; Gainetdinov et al. 2002) and offer new possibilities for screening future drugs.

These new technologies have multiple applications in all domains of the neurosciences. Several of the most spectacular or the most likely to produce medical progress will be presented here. Nonetheless these applications concern other domains and in general all aspects of basic research (for example, the study of neural development; Blackshaw and Livesey 2002).

## Genes and Neuropsychiatric Diseases

More diseases affect  the nervous system than any other human organ system. Accordingly, neurogenetics will benefit enormously from the use of the human genome data. Research on the genetics of neuropsychiatric conditions began well before the decoding of the human genome, and several genes involved in these diseases have been identified. Since the pioneering work on Huntington's disease (Gusella et al. 1996), several other genes have been identified (McGuffin et al. 2001). Some of these diseases are autosomal or X-linked disorders with simple and well-identified determinism (Fisch 2003), but in most cases the determinism of these diseases appears complex. The identification of purely genetic forms is relatively rare; cases of susceptibility genes that predispose the carrier to the disease that might not necessarily develop are more common, as are cases for which no gene contribution has not fully been identified. Alzheimer disease's is, in this regard, exemplary. Three genes clearly linked to its early onset have been identified to date: that for the β-amyloid precursor protein (APP), located on chromosome 21, and those of the PS1 and PS2 presenilins, located on chromosomes 14 and 1. Another gene, for apolipoprotein E, plays a role in more cases, since the presence of its ε4 allele constitutes (with age) the clearest risk factor for Alzheimer's disease. This gene is frequently involved in Alzheimer's disease  but confers predisposition: one can carry the ε4 allele and never develop the disease (even living to a ripe old age; on the other hand, the presence of this allele in its homozygous form makes the onset of Alzheimer's disease very probable). In addition to these genes, the involvement of several other loci is suspected. The genetic determinism of this disease is thus complex, as it appears to be the case for most other neurodegenerative diseases, including: Parkinson's disease, fronto-temporal dementia, prion diseases, amyotrophic lateral sclerosis, etc. In all these cases, rare genetic forms have been identified. Interestingly, the genes concerned code for proteins that tend to aggregate in the nervous system, and whose these aggregates constitute the essential histopathologic characteristic of the disease: senile plaques (principally composed of β-amyloid) in Alzheimer's disease, Lewy bodies (mainly α-synuclein) in Parkinson's disease, Pick bodies (tau protein) in Pick's disease, etc. (Taylor et al. 2002). This research  thus leads not simply to the identification of genes but also to the study of their function and their involvement in the disease. This process occured in Alzheimer's disease: identification of the presenilins by genetic analysis made it possible to study their biological functions (Sisodia and St George-Hyslop 2002). This research has already improved our knowledge of its pathogenesis, and we can hope that this knowledge will lead, in the future, to a renewal of approaches to its treatment. Many neurological diseases are related to aging and can be classified as neurodegenerative disorders. Several others are developmental diseases, like developmental dyslexia which also has a complex genetic basis (Fisher and DeFries 2002). Alterations in development could also be part of schizophrenia aud autism.

Although there is no doubt that genetic factors are involved in the onset of the major psychiatric diseases – autism, schizophrenia and depression – (Hyman 2000; Pfaff et al. 2000; Hyman and Moldin 2001), no gene has been fully accepted as being implicated in the disease (Kelsoe et al. 2001; Levinson et al. 2002). Recently, specific genes have been discovered that influence susceptibility to schizophrenia (Cloninger 2002). Even a disease such as addiction in which clearly environmental factors intervene has a strong genetic component: approximately 40–60% of the risk for alcohol, cocaine, and opiate addiction appears to be genetic (Nestler 2001). Interestingly, in some instances the effect of the gene could be more readily obvious when the environment is taken into account. One example has been described by Caspi et al. (2002), who found that maltreated children with a genotype conferring high levels of monoamine oxidase A (MAOA) expression were less likely to develop antisocial problems.

The success of this search for genes whose mutations are involved in disease comes from two distinct approaches: the study of candidate genes and "blind" searching which involves genome-wide scans. The first approach has found mutations in the genes coding for APP (in Alzheimer's disease) or tau (in fronto-temporal dementia with Parkinson's disease). The second has led to the identification of the parkin (involved in Parkinson's disease), the huntingtin (Huntington's disease) and the presenilins (Alzheimer's disease).

The study of changes in gene expression also offers new perspectives, independently of the study of the genetic mutations involved in these diseases. In their recent studies with cDNA microarrays, Mirnics et al. (this volume) uncovered three types of altered gene expression patterns in the prefrontal cortex of individuals with schizophrenia, compared with matched controls. In light of these findings, Mirnics et al. (2001) have proposed a model according to which different genetic insults related to proteins involved in synaptic communication, combined with pre- or perinatal environmental factors, may lead to the shared clinical manifestations of schizophrenia.

It is now clear that such common diseases as schizophrenia, depression, and neurodegenerative diseases are "complex" or "multifactorial" diseases, that is, they cannot be fully ascribed to mutations in a single gene or to a single environmental factor. New and multivariate strategies are thus needed for genetic research (Phillips and Belknap 2002). Because a new approach such as this may integrate all of these complex diseases, it must take into account other data. These include studies of neurotransmitter (dopamine, glutamate and serotonin) variations, receptors and transporters as candidate genes (Blakely 2001; Buhot 1997; Caspi et al. 2002; Catalano 1999; Hariri et al. 2002; Lakatos et al. 2001; Nestler and Landsman 2001; Wolf et al. 1996), pharmacological studies, brain imaging analysis (Hariri et al. 2002) as well as genetic research  in humans and the development of transgenic animal models. Only this integrated approach can help to solve the puzzle of complex diseases (Sawa and Snyder 2002).

## Genes and Cognitive Faculties

Although the genetic approach in neuropsychiatry raises numerous problems, especially because of the complexity of the factors involved, it at least has the advantage of involving a relatively simple characteristic: the presence or absence of a disease. The study of normal cognitive faculties is thus more complex in many ways. The classic studies of twins and adoptive children have demonstrated the role of genetic factors. According to Hamer (2002), "the effects of heredity [in every aspect of human personality, temperament, cognitive style] were subbstantial, typically representing 30 to 70 % of total variation, and highly replicable across societies and cultures." Nonetheless, these genetic factors remain difficult to identify, nonetheless, because single genes do not determine most human behaviors which usually depend on the interplay between environmental factors and multiple genes (McGuffin et al. 2001). The data and techniques engendered by the study of the human genome should help to identify the precise genes and the corresponding proteins involved – perhaps indirectly – in these behaviors. We could then, as in disease studies, hypothesize biochemical causes of these mental mechanisms. Plomin (this volume) postulates that the two levels of analysis of the gene-behavior pathway, the bottom-up strategy (i.e., functional genomics, how genes work) and the top-down strategy (behavioral genomics) will eventually meet in the brain.

The clearest data about the factors involved in the development of intelligence relate to mental retardation [defined as an overall intelligence quotient – (IQ) – lower than 70], especially that linked to the X chromosome, but it is difficult to know to what extent they illuminate the determinism of cognitive faculties in normal healthy people. Several syndromes of this type and several genes responsible for them have been described. Chelly and Mandel (this volume) found that three important syndromic forms of X-linked mental retardation involve proteins that play a role in chromatin remodeling. This finding provides important insights into the molecular and cellular defects underlying mental retardation. Several genes involved in X-linked mental retardation have been identified (Chelly 1999; Chelly and Mandel, this volume; Meloni et al. 2002; Vervoort et al. 2002).

As of today, few specific genes or even simple markers related to cognitive faculties have been identified with any certainty, although a list of more than 150 candidate genes can be considered (Morley and Montgomery 2001) and a genome-wide scan of 1842 DNA markers for allelic associations with general cognitive ability has been published (Plomin et al. 2001). On the other hand, Lai et al. (2001; see also Fisher, this volume) found that a gene located on chromosome 7, the *FOXP2* gene, is involved in some form of specific language impairment in the KE family. Pinker (2001) asserts that the discovery of this gene is one of the first fruits of the Human Genome Project. This disease is caused by disruption of the gene by a chromosome translocation in the affected members of the KE family. Lai et al. (2001) also found that *FOXP2*, which belongs to a family of genes that encode transcription factors, may have a causal role in the development of the normal brain circuitry that underlies language and speech.

This approach to cognition is only one aspect of the involvement of genetics in psychology. Many reports describe studies of genetic aspects of human and animal behaviors (Edwards and Kravitz 1997; Nelson and Chiavegatto 2001; Pitkow et al. 2001; Sokolowski 2001; Ben-Shahar et al. 2002; Krieger and Ross 2002).

## Genes and Cerebral Development

The discovery of the involvement of a gene such as *FOXP2* in the development of language and the brain illustrates how genetics can help our understanding of cerebral development. Today, methods unrelated to genomics – imaging in particular – are powerful tools for studying this type of genetic involvement in brain development. Using magnetic resonance imaging, Thompson et al. (2001) constructed three-dimensional maps of the brain of dizygotic and monozygotic twins. These maps revealed how brain structure is influenced by genetic differences: genetic factors appear to exert a significant influence on cortical structure in the Broca and Wernicke language areas. Their data also suggest that differences in frontal gray matter may be linked to the 'g' factor (general factor) defined by Charles Spearman as a single factor that underlies the positive correlations among different tests of intelligence. These data suggest that 'g' – a controversial concept even if it is often considered as the primary core of mental ability – is not simply a statistical abstraction but has a biological substrate (Plomin and Kosslyn 2001).

This type of study therefore helps us to understand simultaneously the genetic determinants of cerebral architecture and the origin of individual differences. The nature of these differences is quantitative, as Thompson et al. (2001) suggested and as was underscored by the discovery of an association between total brain volume and intelligence, with substantial genetic mediation (Pennington et al. 2000). However, these differences may also be qualitative, since different individuals probably use different mental strategies to perform the same cognitive task. A gene involved in cerebral cortical size has been recently identified by Bond et al. (2002). A mutation of this gene, called *ASPM*, is associated with autosomal recessive primary microcephaly. This gene is essential for mitotic spindle activity in neuronal progenitor cells (Bond et al. 2002).

## Phylogenesis and Cognitive Differences Between Species

The human genome was decoded at approximately the same time as that of many other species – microbial, plant and animal. Other genomes are currently being decoded. This work is essential to the study of the human genome, since the comparison between genomes makes it possible to individualize the genes and helps to understand the transition between human and other primates.

The current knowledge available from this work so far is that humans do not possess a genetic stock any more extensive than other species. The number of base-pairs contained in our genome is estimated at 3.2 billion. This number is more than that in the genome of the carp (1.7 million), the chicken (1.2 billion), the fly (0.9 million) or the tomato (0.655 million), and certainly more than in the genome of the microorganisms. But it is rather less than in the genome of the mouse (3.454 billion), the trumpet lily (90 billion) or the *Amoeba dubia* (670 billion; Knight 2002). Our species thus does not benefit from any obvious genomic advantage. This observation is not surprising, to the extent that all species currently in existence have, in some way, the same phylogenic seniority, and therefore the same complexity. The difference depends on the type of eco-logical niche they occupy, not on a greater or less degree of evolution. According to Hardy (2001) "this discovery will mark another blow to the anthropocentric view of the universe, similar to the Copernican discovery of heliocentricity. Im-plicit in much of our thought and discussion is the view that man is the pinnacle of evolution and that other species are somehow "failed" humans – neither as intelligent nor as complex as we are." Following Hardy, "this is nonsense of course: they are as well adapted to their environments as we were to roaming the African savannah".

In addition, the evolution of species has included a series of duplications, the role of which was postulated by Susumu Ohno (1970), and of deletions, whose effects are difficult to quantify and which do not express, in any case, differences in organic complexity (Knight 2002).

Nonetheless, because the great complexity of the operation of cognitive facul-ties seems to involve a substantial amount of genetic machinery, we can only wonder about the reason why there is no quantitative difference between mice and humans and why the difference is so small between humans and the fly or the *Caenorhabditis elegans* worm, which possess, respectively, 13,600 and 18,400 genes but only 250,000 and 302 neurons (compared with 10 to 100 billion in humans; that is, roughly a million billion synapses). During the conference, Jean-Pierre Changeux (2001; cf. also Changeux 2002) hypothesized that "the nonlinear revolution of brain complexity in mammals may plausibly be accounted for by nonlinear mechanisms taking place within an evolving network of transcription factors, the genomic revolution taking place primarily at the level of gene regulatory sequences rather than at the level of the structural genes themselves." It is quite possible that the principal differences between humans and apes have less to do with genetic sequences (in the coding genes) than with gene expression, particularly in the brain (Enard et al. 2001).

Most experts today doubt that we can find any major differences in the coding part of the genome between humans and other mammalian species, especially apes. Comparing the data for *Homo sapiens* and apes lets us see if some genes, in particular those involved in cognitive faculties, might differentiate us from our closest relatives in nature whose genetic identity with us is between 95 and 99% (Hoyer et al. 1972; Fujiyama et al. 2002; Enard et al. 2001; Britten et al. 2002). Recent studies of *FOXP2*, a gene involved in speech and language, are particularly interesting in this regard since language is supposed to be a uniquely human trait involved in the development of human culture. Enard et al (2002) recently sequenced the genes that encode the FOXP2 protein in the chimpanzee, gorilla, orangutan, rhesus macaque and mouse. They found that human *FOXP2* contains changes in amino-acid sequences and a pattern of polymorphism suggesting that the gene has been the target of selection during recent human evolution. The fixation of the human allele could be concomitant with the emergence of modern humans during the last 200,000 years of human history.

Another gene of interest is the gene for N-glycolylneuraminic acid (Neu5Gc), a common mammalian sialic acid present in chimpanzees but not humans (Chou et al. 2002). According to Chou et al. (2002), the mutation associated with the inactivation of this gene occurred between 2.5 million and 3 million years ago, after hominids are thought to have stood upright but just before their brain started increasing in size. This mutation could thus have played a part in the evolution of the human brain.

Another comparative approach, more ambitious and complex, consists of seeking the genes involved in the mouse equivalent of g factor, thus benefiting simultaneously from a possible comparison with the human species and the experimental possibilities that this animal species offers (Galsworthy et al. 2002; Plomin 2001), including that of linking neurophysiology with cognition (Nguyen and Gerlai 2002).

## Towards the Decoding of the Cellular Bases of Memory

The tools of biochemistry, genomics and proteomics can be applied to diverse species to enable us to examine the physiological processes involved in memory in a new way. Tang et al. (1999) have shown that overexpression of NMDA receptor 2B in the forebrains of transgenic mice leads to enhanced activation of the NMDA receptors and superior learning and memory ability. This observation led these authors to propose the provocative hypothesis according to which "genetic enhancement of mental and cognition attributes such as intelligence and memory in mammals is feasible" (Tang et al. 1999). This possibility of genetic manipulation reminds us, in a different domain, of that accomplished by Pitkow et al. (2001), that is to say, increased bonding behavior in monogamous male prairie voles by transferring a receptor gene for arginine vasopressin.

Kandel (2001) described a dialogue between genes and synapses. Recent studies in *Aplysia*, *Drosophila* and mice have revealed the interesting possibility that the distinction between short- and long-term memory, which is evident at the behavioral level, results from a fundamental distinction at the cellular level. According to Giustetto et al. (this volume) "the study of long-term memory reveals a ready dialogue between the individual synapse and the nucleus and between the nucleus and the other synapses of the neuron."

On the other hand, Grant and O'Dell (this volume) have tried to focus attention on multiprotein signaling complexes (called hebbosomes) that monitor the patterns of synaptic activity and sort intracellular signals to modify various features of the synapses involved in plasticity and learning.

## The End of Naive Reductionism

Often presented simplistically, the discoveries about the genome and the transcriptome reveal a very high level of complexity that is difficult to decode. The 35,000 genes of the human genome can produce from 500,000 to 700,000 different proteins, and their combinations represent an even wider variety. Moreover, it has recently been discovered that these proteins have an unsuspected "social life:" they sometimes consort with hundreds of other proteins to form very large clusters. Therefore, a multitude of proteins may well be affected when working on a single target (Gavin et al. 2002; Ho et al. 2002). This incredible level of complexity has been clearly shown in yeast. Imagine a situation still more complex at the synapse, where multifunctional complexes associate numerous proteins, representing a considerable level of complexity that is still relatively elementary until we place it within the extremely complex network of synapses and neurons that modern techniques of functional proteomics allow us to approach (Grant and Blackstock 2001). We are, in fact, seeing living the end of naive reductionism (Bloom 2001). Awareness of the extent of this complexity should nonetheless not lead us into exaggerated pessimism about the possibility of progress, especially insofar as the relatively simplistic approach used so far has led to numerous discoveries.

## Treatment Perspectives

In the neurosciences as in other areas of biomedicine, therapeutic applications unfortunately do not immediately follow advances in basic research. The future in this area cannot be predicted with any certainty. It is nonetheless reasonable to imagine advances in the areas of diagnosis and of targeting drugs according to individual genetic differences and types of treatment.

The identification of the genes involved has obvious consequences for postnatal and even prenatal diagnosis. The discovery of mutations in specific genes underlying neurological disorders has led directly to powerful DNA-based diagnostic tests (Bird and Bennett 2000). Beyond applications for genetic counseling, the clinical impact also concerns treatment once it is possible to predict the onset of a disease, before its first symptoms, that is, at a stage when possible preventive therapy may be most effective (for example, antioxidant substances, since oxidative processes seem to be strongly involved in neurodegenerative diseases; Christen 2000).

Pharmacogenomics is another, almost virgin field, for the medical approach. The discovery of susceptibility genes, such as apolipoprotein E, underlines the importance of human polymorphisms in the medical domain. A number of observations suggest that these susceptibility genes may explain differences in the propensity to contract a disease but also in the efficacy of treatment. It is probable that drug efficacy, as well as the extent of drug side effects, is a function of the individual genotype (Catalano 1999; Roses 2000a, b). According to Roses (2000a) "the application of pharmacogenetics to the delivery of medicines will maximize the value of each medicine. Medicines would be prescribed only to those patients where a high probability of efficacy without significant adverse events is expected."

Finally, we must not forget that the ultimate goal of neurogenetics is to treat and prevent neurological disease. We can hope that the discoveries of the genomic and postgenomic era will lead to the development of new treatment strategies: drugs targeted at identified gene mechanisms (secretases in Alzheimer disease's, for example) or gene therapy, sometimes described as the medicine of the 21st century (Verma, this volume). The first clinical trials have shown the difficulty of moving from dream to medical practice but also the possibility of success, since Alain Fischer and colleagues have successfully treated three young babies who suffer from a fatal form of severe combined immunodeficiency syndrome (Cavazanno-Calvo et al. 2000). Numerous projects related to the nervous system are theoretically quite advanced (Barkats et al. this volume). Better knowledge of vectors (Verma, this volume) as well as the possibility of using neural stem cells (Gage 2000; Rossi and Cattaneo 2002) justify the hope for efficacious strategies, which still must prove themselves concretely. Some animal experiments look promising, for example the subthalamic administration of adeno-associated viral vector containing the gene for glutamic acid decarboxylase in a Parkinson's disease rat model, which resulted in strong neuroprotection (Luo et al. 2002)

## A New Field of Ethical Thought

It is self-evident that these new perspectives raise ethical questions, since they involve all the issues of genetic counseling, prenatal diagnosis, the use of embryos to obtain stem cells, the identification of risk factors in the absence of real treatment possibilities, the possible identification by insurance companies of subjects carrying serious diseases, the perspectives of genetic modifications, including of the germ line, cloning, the importance of genetic factors in individual differences, the importance of environmental factors, etc. (Sram et al. 1999), without even mentioning the possible behavior modifications by gene transfer (an eventuality that some animal experiments let us glimpse; Tang et al. 1991; Pitkow et al. 2001). Responding to these questions in a way that will satisfy everyone is obviously impossible, especially as bioethical analyses have shown clearly that diverse religions and forms of thought as well as diverse nations differ on these issues, sometimes radically (Wertz 1991). While it is reasonable to wonder about these consequences of scientific progress, we must not fall into the temptation of a radical criticism of science that will encourage modern obscurantism.

# References

Ben-Shahar Y, Robichon A, Sokolowski MB, Robinson GE (2002) Influence of gene action across different time scales on behavior. Science 296:741–744.

Bird TD, Bennett RL (2000) Genetic counselling and DNA testing. In: Pulst S (ed) Neurogenetics. Oxford University Press, New York, pp.433–442.

Blackshaw S, Livesey R (2002) Applying genomics technologies to neural development. Curr Opin Neurobiol 12:110–114.

Blakely RD (2001) Physiological genomics of antidepressant targets: keeping the periphery in mind. J Neurosci 21:8319–8323.

Bloom FE (2001) What does it all mean to you? J Neurosci 21:8304–8305.

Bond J, Roberts E, Mochida GH, Hampshire DJ, Scott S, Askham JM, Springell K, Mahadevan M, Crow YJ, Markham AF, Walsh CA, Woods CG (2002) ASPM is a major determinant of cerebral cortical size. Nature Genet 32:316–320.

Britten RJ (2002) Divergence between samples of chimpanzee and human DNA sequences is 5% counting indels. Proc Natl Acad Sci USA 21:13633–13635.

Buhot M-C (1997) Serotonin receptors in cognitive behaviors. Current Opin Neurobiol 7:243–254.

Caspi A, McClay J, Moffitt TE, Mill J, Martin J, Craig IW, Taylor A, Poulton R (2002) Role of genotype in the cycle of violence in maltreated children. Science 297:851–854.

Catalano M (1999) Psychiatric genetic '99. The challenge of psychopharmacogenetics. Am J Human Genet 65:606–610.

Cavazzano-Calvo M, Hacein-Bey S, de Saint Basile G, Gross F, Yvon E, Nusbaum P, Selz F, Hue C, Certain S, Casanova J-L, Bousso P, Le Deist F, Fischer A (2000) Gene therapy of human severe combined immunodeficiency (SCID)-X1 disease. Science 288:669–672.

Changeux JP (2001) The non linear evolution of brain-genome complexity from mouse to man. Colloque Médecine et Recherche of the Fondation Ipsen: Neurosciences in the post-genomic era. Paris, december 3, 2001, abstract.

Changeux JP (2002) L'homme de vérité. Odile Jacob, Paris, pp.235–279

Chelly J (1999) Breakthroughs in molecular and cellular mechanisms underlying X-linked mental retardation. Human Mol Genet 81:1833–1838.

Chou H-H, Hayakawa T, Diea S, Krings M, Indriati E, Leakey M, Pääbo S, Satta Y, Takahata N, Varki A (2002) Inactivation of CMP-N-acetylneuraminic acid hydroxylase occurred prior to brain expansion during human evolution. Proc Natl Acad Sci USA 99:11736–11741.

Christen Y (2000) Oxidative stress and Alzheimer's disease. Am J Clin Nutr 71:621S–629S.

Claverie J-M (2001) What if there are only 30,000 human genes? Science 291:1255–1257.

Cloninger CR (2002) The discovery of susceptibility genes for mental disorders. Proc Nat Acad Sci USA 99:13365–13367.

Eberwine J, Kacharmina JE, Andrews C, Miyashiro K, McIntosh T, Becker K, Barrett T, Hinkle D, Dent G, Marciano P (2001) mRNA expression analysis of tissue sections and single cells. J Neurosci 21:8310–8314.

Edwards DH, Kravitz EA (1997) Serotonin, social status and agression. Current Op in Neurobiol 7:812–819.

Enard W, Khaitovitch P, Klose J, Zöllner S, Heissig F, Glavalisco P, Nieselt-Struwe K, Muchmore E, Varki A, Ravid R, Doxiadis GM, Bontrop RE, Pääbo S (2001) Intra- and interspecific variation in primate gene expression patterns. Science 296:340–343.

Enard W, Przeworski M, Fisher SE, Lai CSL, Wiebe V, Kitano T, Monaco AP, Pääbo S (2002) Molecular evolution of *FOXP2*, e gene involved in speech and language. Nature 418:869–872.

Gage FH (2000) Mammalian neural stem cells. Science 287:1433–1438.

Gainetdinov RR, Sotnikova TD, Caron MG (2002) Monoamine transporter pharmacology and mutant mice. Trands Pharmacol Sci 23:367–373.

Galsworthy MJ, Paya-Cano JL, Monleon S, Plomin R (2002) Evidence for general cognitive ability (g) in heterogeneous stock mice and an analysis of potential confounds. Genes Brain Behav 1:88–95.

Gavin, A.-C., Bösche, M., Krause, R., Grandi, P., Marzioch, M., Bauer, A., Schultz, J., Rick, J.M., Michon, A.-M., Cruciat, C.-M., Remor, M., Höfert, C., Schelder, M., Brajenovic, M., Ruffner, H., Merino, A., Klein, K., Hudak, M., Dickson, D., Rudi, T., Gnau, V., Bauch, A., Bastuck, S., Huhse, B., Leutwein, C., Heurtier, M.-A., Copley, R.R., Edelmann, A., Querfurth, E., Rybin, V., Drewes, G., Raida, M., Bouwmeester, T., Bork, P., Seraphin, B., Kuster, B., Neubauer, G. and Superti-Furga, G., Functional organization of the yeast proteome by systematic analysis of protein complexes. *Nature*, 2002, **415**:141–147.

Grant SGN, Blackstock WP (2001) Proteomics in neuroscience: from protein to network. J Neurosci 21:8315–8318.

Gusella JF, MacDonald ME (2000) Molecular genetics: unmasking polyglutamine triggers in neurodegenerative disease. Nature Rev Neurosci 1:109–115.

Gusella JF, McNeil S, Persichetti F, Srinidhi J, Novelleto A, Bird E, Faber P, Vousattel JP, Myerr RH, MacDonald ME (1996) Huntington's disease. Cold Spring Harb Symp Quant Biol New Yyork.

Hardy J (2001) The human genome is sequenced. What does it mean and why is it important? Arch Neurol 58:

Hariri AR, Mattay VS, Tessitore A, Kolachana B, Fera F, Goldman D, Egan MF, Weinberger DR (2002) Serotonin transporter genetic variation and the response of the human amygdala. Science 297:400–403.

Ho, Y., Gruhler, A., Heilbut, A., Bader, G.D., Moore, L., Adams, S.-L., Millar, A., Taylor, P., Bennett, K., Boutilier, K., Yang, L., Wolting, C., Donaldson, I., Schandorfff, R., Shewnarane, J., Vo, M., Taggart, J., Goudreault, M., Muskat, B., Alfarano, C., Dewar, D., Lin, Z., Michalickova, K., Willems, A.R., Sassi, H., Nielsen, P.A., Rasmussen, K.J., Andersen, J.R., Johansen, L.E., Hansen, L.H., Jespersen, H., Podtelejnikov, A., Nielsen, E., Crawford, J., Poulsen, V., Gleeson, B., Pawson, T., Moran, M.F., Durocher, D., Mann, M., Hogue, W.V., Figeys, D. and Tyers, M., Systematic identification of protein complexes in Saccharomyces cerevisiae by mass spectrometry. *Nature*, 2002, **415**:180–183.

Hyman SE (2000) The genetics of mental illlness: implications for practice. Bull World Health Organ 78:455–463.

Hyman SE, Moldin SO (2001) Genetic science and depression: implications for research and treatment. In: The treatment of major depression: bridging the 21st century. American Psychiatric Press, Washington, ppp 83–103.

International Human Genome Sequencing Consortium (2001) Initial sequencing and analysis of the human genome. Nature 409: 860–921.

Kandel ER (2001) The molecular biology of memory storage: a dialogue between genes and synapses. Science 294:1030–1038.

Kelsoe JR, (2002) A genome survey indicates a possible susceptibility locus for bipolar disorder on chromosome 22. Proc Natl Acad Sci USA 98:585–590.

Knight J (2002) All genomes great and small. Nature 417:374–376.

Krieger MJB, Ross KG (2002) Identification of a major gene regulating complex social behavior. Science 295:328–332.

Lai CSL, Fisher SE, Hurst JA, Vargha-Khadem F, Monaco AP (2001) A forkhead-domain gene is mutated in a severe speech and language disorder. Nature 413:519–523.

Lakatos K, Nemoda Z, Toth I, Ronai Z, Ney K, Sasvari-Szekely M, Gervai J (2001) Further evidence for the role of the dopamine D4 receptor (DRD4) gene in attachment disorganization: interaction of the exon III 48 bp repeat and the D521 C/T promoter polymorphisms. Mol Psychiat 7:27–31.

Levinson DF, Holmans PA, Laurent C, Riley B, Pulver AE, Gejman PV, Schwab SG, Williams NM, Owen MJ, Wildenauer DB, Sanders AR, Nestadt G, Mowry BJ, Wormley B, Bauché S, Soubigou S, Ribble R, Nertney DA, Liang KY, Martinolich L, Maier W, Norton N, Williams H, Albus M, Carpenter EB deMarchi N, Ewen-White KR, Walsh D, Jay M, Deleuze J-F, O'Neill FA, Papadimitriou G, Weilbaecher A, Lerer B, O'Donovan MC, Dikeos D, Silverman JR, Kendler KS, Mallet J, Crowe RR, Walters M (2002) No major schizophrenia locus detected on chromosome 1q in a large multicenter sample. Science 296:739–741.

McGuffin P, Riley B, Plomin R (2001) Toward behavioral genomics. Science 291:1248–1249.

Meloni I, Muscettola M, Raynaud M, Longo I, Bruttini M, Moizard M-P, Gomot M, Chelly J, des Portes V, Fryns J-P, Ropers H-H, Magi B, Bellan C, Volpi N, Yntema HG, Lewis SE, Schaffer JE, Renieri A (2002) FACL4, encoding fatty acid-CoA ligase 4, is mutated in nonspecific X-linked mental retardation. Nature Genet 30:436–440.

Mirnics K, Middleton FA, lewis DA, Levitt P (2001) Analysis of complex brain disorders with microarrays: schizophrenia as a disease of the synapse. Trends Neurosci 24:479–486.

Morley KI, Montgomery GW (2001) The genetics of cognitive processes: candidate genes in humans and animals. Behav Genet 31:511–531.

Nelson RJ, Chiavegatto S (2001) Molecular basis of aggression. Trends Neurosci 12:713–719.

Nestler EJ (2001) Psychogenomics: opportunities for understanding addiction. J Neurosci 21:8324–8327.

Nestler EJ, Landsman D (2001) Learning about addiction from the genome. Nature 409:834–835.

Nguyen PV, Gerlai R (2002) Behavioral and physiological characterization of inbred mouse strains: prospects for elucidating the molecular mechanisms of mammalian learning and memory. Genes Brain Behav 1:72–81.

Pääbo S (2001) The human genome and our view of ourselves. Science 291:1219–1220.

Pennington BF, Filipek PA, Lefly D, Chhabildas N, Kennedy DN, Simon JH, Filley CM, Galaburda A, DeFries JC (2000) A twin MRI study of size variations in the human brain. J Cogn Neurosci 12:223–232.

Pfaff DW, Berrettini WH, Joh TH, Maxson SC, eds. (2000) Genetic influences on neural and behavioral functions. CRC Press, Boca Ration, Fl.

Phillips TJ, Belknap JK (2002) Complex-trait genetics: emergence of multivariate strategies. Nature Rev Neurosci 3:478–485.

Pinker S (2001) Talk to genetics and vice versa. Nature 413:465–466.

Pitkow LJ, Sharer CA, Ren X, Insel TR, Terwilliger EF, Young (2001) Facilitation of affiliation and pair-bond formation by vasopressin receptor gene transfer into the ventral forebrain of a monogamous vole. J Neurosci 15:7392–7396.

Plomin R (2001) The genetics of g in human and mouse. Nature Rev Neurosci 2:136–141.

Plomin R, Kosslyn SM (2001) Genes, brain and cognition. Nature Neurosci 4:1153–1155.

Plomin R , Hill L, Craig IW, McGuffin P, Purcell S, Sham P, Lubinski D, Thompson LA, Fisher PJ, Turic D, Owen MJ (2001) A genome-wide scan of 1842 DNA markers for allelic associations with general cognitive ability: a five-stage design using DNA pooling and extreme selected groups. Behav Genet 31:497–509.

Roses AD (2000a) Pharmacogenetics and the practice of medicine. Nature 405:857–865.

Roses AD (2000b) Pharmacogenetics and the future of drug development and delivery. Lancet 355:1358–1361.

Rossi F, Cattaneo E (2002) Neural stem cell therapy for neurological diseases: dreams and reality. Nature Rev Neurosci 3:401–409.

Sachidanandam R et al. (2001) A map of human genome sequence variation containing 1.42 million single nucleotide polymorphisms. Nature 409:928–933.

Sawa A, Snyder SH (2002) Schizophrenia: diverse approaches to a complex disease. Science 296:692–695.

Sisodia SS, St George-Hyslop PH (2002) γ-secretase, notch, Aβ and Alzheimer's disease: where do the presenilins fit in? Nature Rev Neurosci 3:281–290.

Sokolowski MB (2001) *Drosophila*: genetics meet behaviour. Nature Rev Genet 2:879–890.

Sram RJ, Bulyzhenkov V, Prilipko L, Christen Y., eds. (1991) Ethical issues of molecular genetics in psychiatry. Springer Verlag, Heidelberg.

Tang Y-P, Shimizu E, Dube GR, Rampon C, Kerchner GA, Zhuo M, Liu G, Tsien JZ (1999) Genetic enhancement of learning and memory in mice. Nature 401:63–69.

Taylor JP, Hardy J, Fischbeck KH (2002) Toxic proteins in neurodegenerative disease. Science 296:1991–1995.

Thompson PM., Cannon TD, Narr KL, van Erp T, Poutanen V-P, Huttunen M, Lönnqvist J, Standertskjöld-Nordenstam C-G, Kaprio J, Khaledy M, Dail R, Zoumalan CI, Toga AW (2001) Genetic influences on brain structure. Nature Neurosc 4:1253–1258.

Venter CJ, Adams MD, Myers EW, Li PW, Mural RJ, Sutton GG, Smith HO, Yandell M, Evans CA, Holt RA, Gocayne JD, Amanatides P, Ballew RM, Huson DH , Russo Wortman J, Zhang Q, Kodira CD, Zheng XH, Chen L, Skupski M, Subramanian G, Thomas PD, Zhang J, Gabor Miklos GL, Nelson C, Broder S, Clark AG, Nadeau J, McKusick VA, Zinder N, Levine AJ, Roberts RJ, Simon M, Slayman C, Hunkapiller M, Bolanos R, Delcher A, Dew I, Fasulo D, Flanigan M, Florea L, Halpern A, Hannenhalli S, Kravitz S, Levy S, Mobarry C, Reinert K, Remington K, Abu-Threideh J, Beasley E, Biddick K, Bonazzi V, Brandon R, Cargill M, Chandramouliswaran I, Charlab R, Chaturvedi K, Deng Z, Di Francesco V, Dunn P, Eilbeck K, Evangelista C, Gabrielian AE, Gan W, Ge W, Gong F, Gu Z, Guan P, Heiman TJ, Higgins ME, Ji R-R, Ke Z, Ketchum KA, Lai Z, Lei Y, Li Z, Li J, Liang Y, Lin X, Lu F, Merkulov GV, Milshina N, Moore HM , Naik AK, Narayan VA, Neelam B, Nusskern D, Rusch DB, Salzberg S, Shao W, Shue B, Sun J, Wang ZY, Wang A, Wang X, Wang J, Wei M-H, Wides R, Xiao C, Yan C, Yao A, Ye J, Zhan M, Zhang W, Zhang H, Zhao Q, Zheng L, Zhong F, Zhong W, Zhu SC, Zhao S, Gilbert D, Baumhueter S, Spier G, Carter C, Cravchik A, Woodage T, Ali F, An H, Awe A, Baldwin D, Baden H, Barnstead M, Barrow I, Beeson K, Busam D, Carver A, Center A, Cheng ML, Curry L, Danaher S, Davenport L, Desilets R, Dietz S, Dodson K, Doup L, Ferriera S, Garg N, Gluecksmann A, Hart B, Haynes J, Haynes C, Heiner C, Hladun S, Hostin D, Houck J, Howland T, Ibegwam C, Johnson J, Kalush F, Kline L, Koduru S, Love A, Mann F, May D, McCawley S, McIntosh T, McMullen I, Moy M, Moy L, Murphy B, Nelson K, Pfannkoch C, Pratts E, Puri V, Qureshi H, Reardon M, Rodriguez R, Rogers Y-H, Romblad D, Ruhfel B, Scott R, Sitter C, Smallwood M, Stewart E, Strong R, Suh E, Thomas R, Tint NN, Tse S, Vech C, Wang G, Wetter J, Williams S, Williams M, Windsor S, Winn-Deen E, Wolfe K, Zaveri J, Zaveri K, Abril JF, Guigó R, Campbell MJ, Sjolander KV, Karlak B, Kejariwal A, Mi H, Lazareva B, Hatton T, Narechania A, Diemer K, Muruganujan A, Guo N, Sato S, Bafna V, Istrail S, Lippert R, Schwartz R, Walenz B, Yooseph S, Allen D, Basu A, Baxendale J, Blick L, Caminha M, Carnes-Stine J, Caulk P, Chiang YH, Coyne M, Dahlke C, Deslattes Mays A, Dombroski M, Donnelly M, Ely D, Esparham S, Fosler C, Gire H, Glanowski S, Glasser K, Glodek A, Gorokhov M, Graham K, Gropman B, Harris M, Heil J, Henderson S, Hoover J, Jennings D, Jordan C, Jordan J, Kasha J, Kagan L, Kraft C, Levitsky A, Lewis M, Liu X, Lopez J, Ma D, Majoros W, McDaniel J, Murphy S, Newman M, Nguyen T, Nguyen N, Nodell M, Pan S, Peck J, Peterson M, Rowe W, Sanders R, Scott J, Simpson M, Smith T, Sprague A, Stockwell T, Turner R, Venter E, Wang M, Wen M, Wu D, Wu M, Xia A, Zandieh A, Zhu X (2001) The sequence of the human genome. Science 291: 1304–1351.

Vervoort VS, Beachem MA, Edwards PS, Ladd S, Miller KE, de Mollerat X, Clarkson K, DuPont B, Schwartz CE, Stevenson RE, Boyd E, Srivastava AK (2002) AGTR2 mutations in X-linked mental retardation. Science 296:2401–2403.

Wertz DC (1991) Lessons from an international survey of medical geneticists. In: Ethical issues of molecular genetics in psychiatry (Sram RJ, Bulyzhenkov V, Prilipko L, Christen Y., eds.). Springer Verlag, Heidelberg, pp.61–86.

Wolf SS, Jones DW, Knable MB, Gorey JG, Lee KS, Hyde TM, Coppola R, Weinberger DR (1996) Tourette syndrome: prediction of phenotypic variation in monozygotic twins by caudate nucleus D2 receptor binding. Science 273:1225–1227.

Wong PC, Cai H, Borchelt DR, Price DL (2002) Genetically engineered mouse models of neurode-generative diseases. Nature Neurosci 5:633–639.

Zoghbi HY, Botas J (2002) Mouse and fly models of neurodegeneration. Trends Genet 18:463–471.

# Memory and the Regulation of Chromatin Structure

M. Giustetto[1*], Z. Guan[2*], S. Lomvardas[3], Dimitris Thanos[3], J. H. Schwartz[2], E.R. Kandel[1,2]

One of the striking behavioral observations to emerge from the study of memory storage is that long-term memory – lasting days – differs mechanistically from short-term memory – lasting minutes – in requiring the synthesis of new protein (Davis and Squire, 1984). Complementary studies of simple implicit memory storage in *Aplysia* and more complex explicit memory storage in mice have revealed that these two temporally and mechanistically different phases of behavioral memory are reflected in temporally and mechanistically distinct phases of synaptic plasticity in the neurons that participate in memory storage. Within these neurons, a core signaling pathway has been identified that is conserved from mollusks to mice and that is critical both for the conversion of short- to long-term synaptic plasticity and for the conversion of short- to long-term memory. This core pathway involves the recruitment by cAMP, PKA, and p42 MAPK, the translocation of PKA and p42 MAPK to the nucleus leading to the activation of a gene cascade beginning with the removal of the repression mediated by CREB-2 (ATF4), and the activation of CREB-1. CREB-1, in turn, induces the CAAT box enhancer binding protein (C/EBP) and other immediate response genes, leading to the growth of new synaptic connections (Kandel, 2001).

The finding that long-term plasticity and long-term memory recruit constitutively expressed transcription factors in the nucleus that induce immediate response genes provided an explanation for why protein synthesis during and immediately after learning is important for the activation of long-term memory. However, this finding also raised a deep question in the cell biology of memory storage: If long-term synaptic plasticity involves transcription and therefore the nucleus – an organelle shared by all the synapses of a neuron – does this mean that the unit of long-term information storage in the brain is the whole cell? Or can the induced gene products be compartmentalized spatially or functionally, so that they selectively alter the function of some synapses and not others?

[1] Howard Hughes Medical Institute,. 1051 Riverside Drive, New York, New York 10032
[2] Center for Neurobiology and Behavior, College of Physicians and Surgeons, Columbia University, New York State Psychiatric Institute, 1051 Riverside Drive, New York, New York 10032
[3] Department of Biochemistry & Molecular Biophysics, College of Physicians and Surgeons, Columbia University, 630 West 168th Street, New York, New York 10032
* These two authors contributed equally to this work.

Mallet/Christen
Neurosciencess at the Postgenomic Era
© Springer-Verlag Berlin Heidelberg 2003

This question has been addressed on the cellular level in a culture system consisting of a single sensory neuron with a bifurcated axon that makes synaptic contact with two spatially separated, target motor neurons (Martin et al., 1997). In this culture system, five pulses of serotonin (5-HT), a facilitatory neurotransmitter released by sensitizing stimuli to the tail, produces a synapse-specific, long-term facilitation that persists more than 72 hours, requires CREB-1-mediated transcription, and leads to the growth of new synaptic contacts. However, despite its evident synapse specificity, the induction of synapse-specific and transcription-dependent, long-term facilitation also reduces dramatically the threshold for long-term facilitation cell-wide Infact, so that during a critical time period of about one to two hours, the long-term process can be "captured" and new connections grown at another set of synaptic connections with only a single marking pulse of 5-HT, which normally produces facilitation that lasts only minutes. Thus, rather than being rigorously synapse-specific as is the short-term process, the long-term process exerts a subthreshold influence that is cell-wide, so that the modulation of synaptic transmission at one set of terminals can influence transmission at a distant set of terminals through an action on CREB-1 in the nucleus.

These experiments on synapse-specific facilitation and synaptic capture first illustrated that the logic for long-term facilitation differs from short-term facilitation, and opened up for study a new feature of neuronal function – long-term, long distance neuronal integration. We have more gone on to ask two further questions:

1) How general is this mechanism of integration? Does it also apply to inhibition? What happens when inhibition and facilitation compete with one another?
   and
2) if synapse-specific inhibition and facilitation occur, where does their integration take place?

In earlier work we found that FMRFamide, an enkephalin-related peptide, produces both a short- and long-term, presynaptic inhibitory action in the connections between the sensory and motor neurons that is opposite that of 5-HT (Montarolo et al. 1988). This inhibition is mediated through a receptor coupled to p38 MAP kinase that activates phospholipase A2 and liberates arachidonic acid (Piomelli et al. 1987; Guan et al. 2002). We show here that this inhibition also has a long-lasting, synapse-specific form that is transcriptionally dependent. But inhibition is mediated by CREB-2, not by CREB-1. Thus, CREB-2 is not only a repressor of CREB-1 (Bartsch et al. 1995) but is also an activator of genes for long-term inhibition.

Given the fact that both long-term facilitation and inhibition involve transcription, how does a sensory cell integrate opposing long-term signals from FMRFamide and 5-HT applied independently at two different branches? We found that, when five pulses of FMRFamide were applied to one branch and five pulses of 5-HT to the other branch, long-term inhibition was fully expressed but long-term facilitation was completely blocked. Thus, long-term synaptic integration differs dramatically from short-term integration, where one pulse of 5-HT and one pulse of FMRFamide applied to different branches each gave rise to the synapse-specific, short-term processes at the individual branches. Unlike the synapse-specific logic of short-term process, the logic of the long-term process involves the nucleus as well as the synapse, and in the nucleus the long-term inhibition dominates.

How does this come about? What is the molecular logic of the long-term process? One major clue has come from studies of the cell cycle by Chrivia et al. (1993), which showed that to activate transcription, CREB-1 recruits the CREB binding protein (CBP). CBP in turn acetylates specific lysine residues on core histones of the nucleosome to lead to alterations in chromatin structure (Giles et al. 1998; Goodman and Smolik 2000). We therefore asked: is chromatin modification limited to cellular development and differentiation or do the signal transduction pathways recruited for learning also lead to chromatin modification? Is the integrative action between long-term inhibition and facilitation played out at the level of the nucleosome, the building block of chromatin? If so, can we obtain new insights into the molecular biology of the long-term process by examining chromatin structure in detail?

Chromatin is made up of nucleosome cores and linker DNA. The nucleosome core consists of an octomer of histone containing two each of four different types (H2A, H2B, H3, and H4), along with about 150 bp DNA (Kornberg and Lorch 1999). The nucleosomes tend to cluster around regulatory regions of genes where they inhibit transcription. One way to regulate gene expression is to recruit, to the promoter, histone-modifying enzymes that alter chromatin structure. The tight packaging of DNA into chromatin is based on the interaction between negative charge of DNA and positive charge of histone, which derives from positively charged lysine and arginine residues in histones. When lysine residues are acetylated, histone loses its positive charge so that the negatively charged DNA is released from chromatin, enabling other DNA binding proteins to be recruited to DNA. Acetylation of lysine residues tends to correlate with gene transcription and deacetylation with gene repression. This tendency has been particularly well established for H3 and H4 (Kornberg and Lorch 1999). Despite the extensive study of transcription in brain, however, nothing is known about how external events that affect transcription modulate chromosomal structure in neurons, and whether this modulation introduces an additional level of integration into memory-related neuronal function.

To study chromatin structure and protein-DNA interaction in vivo, we used chromatin immunoprecipitation (ChIP; Solomon et al. 1988; Kuo and Allis 1999; Orlando 2000; Guan et al. 2002). Using this assay, we found that the modulatory transmitters 5-HT and FMRFamide produce critical modifications in chromatin that lead to gene activation on the one hand and to gene repression on the other. Thus, in the basal state, CREB resides on the C/EBP promoter without CBP. 5-HT activates CREB1 to recruit CBP to the promoter, and this recruitment correlates with histone acetylation of a specific lysine residue in H4 and the induction of C/EBP transcription in vivo. When FMRFamide is given alone, it blocks the induction of C/EBP by replacing CREB1 with CREB2 at C/EBP promoter. CREB2 recruits histone deacetylases (HDAC5) of the type II variety, which in turn reduce dramatically the basal level of histone acetylation by histone deacetylation, leading to repressing the gene. When FMRFamide and 5-HT were given together, recruitment of CREB-2 and HDAC5 turned off the induction of C/EBP as effectively as if 5-HT were not present. Thus, we provide direct evidence that the signals from facilitatory and inhibitory neurotransmitters important for memory activate signal transduction pathways that extend to the nucleosomes, and that the integrative action for long-term synaptic plasticity is fought on the battleground of chromatin structure. In so doing, our studies provide the initial evidence that modulatory transmitters important for learning modulate chromatin structure bidirectionally (Guan et al. 2002).

## Discussion

Our study of long-term memory reveals a ready dialogue between the individual synapse and the nucleus and between the nucleus and the other synapses of the neuron. As a result of this long-range interaction, a new level of integration is evident with long-term plasticity that is different from that used by the short-term process. Short-term synaptic plasticity is synapse-specific. In the long-term, activity at one terminal can exert a long-range, long-term influence on another with the results that the response of a synapse is not simply determined by its own history of activity, but also by the activity of other synapses. How is this novel, long-term, non-Sherringtonian, signal integration achieved?

A neuron can simultaneously receive excitatory and inhibitory inputs through many spatially separate synapses, and it must integrate these competing inputs before deciding its response. The classical Sherringtonian theory of neuronal integration describes the short-term integration: the neuron summates the inputs at the axon hillock to decide if it will fire an action potential. This short-term integration happens on the cell membrane to regulate the opening of voltage-dependent $Na^+$ channels. Inhibitory inputs dominate this integration because they often end on the cell body, which makes them effective to interrupt and override excitatory inputs.

Our study of long-term memory in *Aplysia* now provides the initial evidence for a novel mechanism for long-term neuronal integration. When an *Aplysia* sensory neuron receives repeated facilitatory (5-HT) and inhibitory (FMRFamide) inputs through spatially separate synapses, it integrates the competing inputs to make its decision for long-term response, and the inhibitory response is dominant cell-wide. This long-term integration happens in the nucleus to bidirectionally regulate chromatin structure and gene induction. Inhibitory inputs dominate in long-term integration by overriding the effect of facilitatory inputs on histone acetylation and gene induction. Thus a neuron integrates opposite inputs at several levels: on the cell membrane to determine short-term response, and in the nucleus to determine long-term response.

# References

Bailey, C.H., Montarolo, P., Chen, M., Kandel, E.R., Schacher, S. (1992). Inhibitors of protein and RNA synthesis block structural changes that accompany long-term heterosynaptic plasticity in Aplysia. Neuron 9, 749–58.

Bartsch D, Ghirardi M, Skehel PA, Karl KA, Herder SP, Chen M, Bailey CH, Kandel ER (1995) *Aplysia* CREB2 represses long-term facilitation: relief of repression converts transient facilitation into long-term functional and structural change. Cell 83: 979–992.

Chrivia JC, Kwok RP, Lamb N, Hagiwara M, Montminy MR, Goodman RH (1993) Phosphorylated CREB binds specifically to the nuclear protein CBP. Nature 365: 855–859.

Davis HP, Squire LR (1984) Protein synthesis and memory: a review. Psychol Bull. 96: 518–559.

Giles RH, Peters DJ, Breuning MH (1998) Conjunction dysfunction: CBP/p300 in human disease. Trends Genet 14: 178–183.

Goodman RH, Smolik S (2000) CBP/p300 in cell growth, transformation, and development. Genes Dev 14: 1553–1577.

Guan Z, Giustetto M, Lomvardas S, Kim JH, Miniaci MC, Schwartz JH, Thanos D, Kandel ER (2002) Integration of long-term-memory-related synaptic plasticity involves bidirectional regulation of gene expression and chromatin structure. Cell 111: 483–493.

Kandel ER (2001) The molecular biology of memory storage: a dialogue between genes and synapses. Science 294: 1030–1038.

Kornberg RD, Lorch Y (1999) Twenty-five years of the nucleosome, fundamental particle of the eukaryote chromosome. Cell 98: 285–294.

Kuo MH, Allis CD (1999) In vivo cross-linking and immunoprecipitation for studying dynamic Protein:DNA associations in a chromatin environment. Methods 19: 425–433.

Martin KC, Casadio A, Zhu H, EY, Rose J, Chen M, Bailey CH, Kandel ER (1997) Synapse-specific, long-term facilitation of *Aplysia* sensory to motor synapses: a function for local protein synthesis in memory storage. Cell 91: 927–938.

Montarolo PG, Kande ER, Schacher S (1988) Long-term heterosynaptic inhibition in *Aplysia*. Nature 333: 171–174.

Orlando V (2000) Mapping chromosomal proteins in vivo by formaldehyde-crosslinked-chromatin immunoprecipitation. Trends Biochem Sci. 25: 99–104.

Piomelli D, Volterra A, Dale N, Siegelbaum SA, Kandel ER, Schwartz JH, Belardetti F (1987) Lipoxygenase metabolites of arachidonic acid as second messengers for presynaptic inhibition of *Aplysia* sensory cells. Nature 328: 38–43.

Solomon MJ, Larsen PL, Varshavsk A (1988) Mapping protein-DNA interactions in vivo with formaldehyde: evidence that histone H4 is retained on a highly transcribed gene. Cell 53: 937–947.

# The Hebbosome Hypothesis of Learning: Signaling Complexes Decode Synaptic Patterns of Activity and Distribute Plasticity

*S.G.N. Grant[1] and Thomas J. O'Dell[2]*

## Summary

Patterns of action potentials are the universal code for information transfer in the nervous system and were first identified as the basis of sensory perception and motor action. Hebb proposed that the mechanism of learning would involve patterns of action potentials driving "metabolic" processes that lead to lasting changes in neurons. It is now widely accepted that these metabolic events involve signal transduction from synaptic neurotransmitter receptors and subsequent changes in a remarkably wide variety of downstream effector mechanisms. Here we present the hypothesis that synapses contain specialized molecular devices that read patterns of action potentials and instruct long-term changes in the properties of the synapse and neuron. These functions are contained within ~2-3 MDa multiprotein complexes found at excitatory synapses, referred to as hebbosomes (a conjunction of Hebb and "some," Greek for body). A key feature of hebbosome signaling complexes is an ability to coordinate multiple pathways downstream from a given receptor and integrate signals from multiple neurotransmitter receptors. This orchestration of signaling provides computation and control of a combination of effector pathways, which then alter the structure and function of the neuron. In this way hebbosomes write the history of synaptic activity into the intracellular molecular networks of the neuron. Consistent with these functions, interference with many different hebbosome proteins impairs learning and other developmental and pathological processes governed by patterned activity in rodents and humans.

## Introduction

This paper provides an overview of a series of studies on the molecular mechanisms of synapse function that lead to the emergence of the hebbosome hypothesis. Here the main objective is to present the reader with the model that synapses contain specialized multiprotein signaling complexes that perform the task of converting patterns of action potentials into cellular changes. The scope of this

[1] Department of Neuroscience, University of Edinburgh, Edinburgh EH9-3JQ, United Kingdom
[2] Department of Physiology and Interdepartmental Ph.D. Program for Neuroscience, UCLA School of Medicine, Los Angeles, California, 90095

Mallet/Christen
Neurosciencess at the Postgenomic Era
© Springer-Verlag Berlin Heidelberg 2003

paper is not to comprehensively review the literature but to set the stage for a discussion of a model with ramifications for basic and clinical neuroscience.

## Historical background

### *Patterns of nerve cell firing and theories of plasticity and learning*

The notion that behavioural experience or learning leads to plastic changes in the structure or function of the brain dates to antiquity (Finger 1994). With the emergence of the neuron doctrine in the late 19th century and the recognition of basic neuronal architecture, it was proposed that "structural changes" in neurons might reflect or enable learning to be encoded. The idea that "resistance" between neurons might alter with learning arose around the same time.

In the early 20th century, electrophysiological studies produced new findings that were used to update these simple models. Sherrington's description of synaptic transmission (1906) and Adrian's (1928) discovery that information is coded in the pattern of action potentials provided a basis for the suggestion that a "reverberatory transient trace" mechanism could be set up in neural circuits and thereby represent a transient "memory" of the stimulus (Hebb 1949). Hebb stated, "there may, then, be a memory trace that is wholly a function of a pattern of neural activity, independent of any structural change;" this was suggested to account for short-term learning that was instantaneously established, such as the temporary recollection of a telephone number.

Hebb further developed this idea of circuits or ensembles of neurons encoding memories within his "dual trace model," in which the transient reverberatory trace might lead to a second trace that provided permanence in the form of "structural change." He said, "if some way can be found of supposing that a reverberatory trace might cooperate with the structural change, and carry the memory until the growth change is made, we should be able to recognize the theoretical value of the trace, which is an activity only, without having to ascribe all memory to it." A component of this model, which has gained more attention than his general model, is the now famous neurophysiological postulate: "Let us assume then that the persistence or repetition of a reverberatory activity (or "trace") tends to induce lasting cellular changes that add to its stability. The assumption can be precisely stated as follows: When an axon of cell A is near enough to excite a cell B and repeatedly or persistently takes part in firing it, some growth process or metabolic change takes place in one or both cells such that As efficiency, as one of the cell firing B, is increased."

Hebb effectively combines elements of preexisting models (patterns of action potentials; synapse resistance; synapse structure) into one where patterns of firing are detected by some mechanism that is linked to cellular machinery for "metabolic" change that alters structure and resistance between neurons. Implicit within this physiological model is the assumption that there could be specific molecular mechanisms for detecting and driving the plastic changes.

These pre-1950 models were proposed on the basis of a comparatively limited knowledge of synaptic physiology and structure, yet have endured primarily because of the intrinsic appeal of the idea that changing synapse strength with activity (synaptic plasticity) would be a suitable way to encode long-term memories. Although enormously persuasive, considerable debate remains amongst neuroscientists as to the causal or correlative relationship between synaptic plasticity and learning.

To obtain a detailed and reductionist molecular account of the processes of plasticity and learning, a variety of animal models has been explored; perhaps the most significant amongst the invertebrates being *Aplysia californica* (a marine sea slug; Kandel 2001) and *Drosophila melanogaster* (the fruit fly; Waddell and Quinn 2001) and amongst the mammals, the rat and mouse. Each of these model organisms has made significant contributions to the current understanding of molecular signaling cascades; however, nowhere has the challenge to identify this molecular machinery attracted a greater number of investigators than the mammalian hippocampus.

### The hippocampus model and long-term potentiation

Within the hippocampus, experimentally induced patterns of action potentials lead to long-lasting plastic changes in the strength of connection between neurons – the phenomenon of long-term potentiation (LTP; Bliss and Lomo 1973). The hippocampus slice preparation, like the frog nerve-muscle preparation before it, has served as an ideal model system for these electrophysiological studies.

LTP has been extensively studied between the axonal projections of CA3 neurons onto dendrites of CA1 neurons (Bliss and Collingridge 1993). At these excitatory synapses, pharmacological antagonists of glutamate receptor subtypes lead to the definition of N-methyl-D-aspartate receptor (NMDAR)-dependent forms of LTP (Collingridge et al. 1983). In NMDAR-dependent forms of LTP, multiple, distinct temporal phases can be found that can be broadly divided into "induction" and "expression." The induction phase involves the first few minutes during and after the stimulus train, and the expression phase the period thereafter. It is the induction phase that involves the detection of the firing pattern and the initial signal transduction changes. This finding is demonstrated by the observations that disruption of the NMDAR and various signal transduction enzymes during this first few minutes interferes with the level of LTP obtained. In contrast, inhibitors of second messengers during the expression phase generally have no effect on synaptic efficacy.

### Molecular genetic studies of LTP and learning

The 1990s was the decade of molecular biology in LTP. Beginning with the cloning of genes for glutamate receptors and the introduction of mouse genetics lead to new ways to test the function of molecules in LTP. In 1992, five knockout mice were reported in LTP studies, including CamKII, Fyn, Src, Yes and Abl (Silva et al. 1992; Grant et al. 1992). CamkII mutant mice showed defects in the induction of LTP consistent with the earlier pharmacological studies (Malinow et al. 1988;

Malenka et al. 1989) and Fyn tyrosine kinase mutant mice showed defects similar to those observed with tyrosine kinase inhibitors (O'Dell et al. 1991). Prior to this experiment there was no reason to implicate Fyn in LTP, thereby demonstrating that gene targeting could be used as a gene discovery tool for LTP. These studies opened the floodgates on mammalian genetics and learning.

Mutations introduced into genes encoding synaptic signaling proteins provide a powerful tool in studying the mechanisms of plasticity and learning. Despite criticisms by some neuroscientists of genetics in these types of applications, mutations have been the essential ingredient in dissection of signal transduction pathways in a vast array of cells and organisms. Discussions on issues of compensation and the utility of mutant mice in identifying steps in pathways can be found elsewhere (Grant et al. 1995). Although mutation combined with biochemical studies and analysis of phenotype is a general method for characterizing pathways, these types of approaches become increasingly powerful when combinations of mutations are studied. Little of this type of genetic study has been applied to LTP, but it is likely to become increasingly important.

The large number of mutant mice that have been shown to have defects in the induction of LTP, combined with the even larger number of pharmacological studies showing various proteins are involved with induction, has lead to an important turning point. It has become apparent that the molecular complexity of LTP and its related form of plasticity, LTD, is very high (Sanes and Lichtman 1999). Classification of these molecules shows roles for neurotransmitter receptors, cell adhesion proteins, adaptor molecules, signaling proteins, proteases, translation and transcription factors and cytoskeletal proteins. When a new protein is reported, there is inevitably a schematic diagram showing NMDAR with a sequence of arrows projecting from it, suggesting putative pathways. An immediate challenge must now be to understand the composition and relationships of postsynaptic signaling networks from the NMDAR and rationalize the induction step of LTP.

It is beyond the scope of this article to systematically review the literature on this large list of molecules or their interactions, but we intend to draw attention to key elements and emerging themes and problems in the field.

## Problems for the synaptic plasticity and learning models emerging from molecular studies

The molecular complexity of plasticity and learning raises two central problems that will be discussed in detail in later parts of this paper:

### Problem 1:
*How can the NMDAR be involved with so many diverse cell biological mechanisms?* (Table 1).

To illustrate this problem, let us consider four of these mechanisms as examples; namely AMPA receptor trafficking, protein synthesis, transcription and cytoskeletal regulation. From existing knowledge of each of these processes, which have been extensively studied in non-neuronal cells, it is apparent that they have

**Table 1.** Diverse roles of the NMDA receptor (NMDAR). The NMDAR has been studied in the context of many aspects of neurobiology using pharmacological, biochemical, genetic and other methods. Listed here are phenomena in which the NMDAR has been shown to be involved. The lists are divided into three main subcategories: 1) cell biological, 2) electrophysiological, and 3) organismal (whole animal). The involvement of the NMDAR in this diverse set of biological functions underpins the need to identify ways in which the NMDAR can be part of multiple intracellular signaling mechanisms, all of which may need to be carefully coordinated. These lists are incomplete but serve to draw attention to the complex organization of signaling.

| NMDA Receptor Signalling | |
| --- | --- |
| Cell Biological | – AMPA phosphorylation<br>– AMPA trafficking<br>– Transcription<br>– mRNA trafficking<br>– mRNA localisation<br>– Protein synthesis<br>– Protein dynamics<br>– Cytoskeletal changes<br>– Presynaptic excitability<br>– Retrograde signalling |
| Electrophysiological | *Synaptic Plasticity*<br>– LTP<br>– LTD<br>– Depotentiation<br>– Metaplasticity<br>– Spike timing plasticity<br><br>*Non-synaptic plasticity*<br>– EPSP-Spike coupling |
| Whole Animal | – Learning<br>– Neuropathic pain<br>– Cerebral ischaemia |

several features in common. Each is regulated by sets of proteins often organized into multiprotein complexes (e.g., ribosomes, mRNA transcription complexes). Second, they are controlled by multiple signal transduction pathways, many involving kinases. Somehow the NMDAR requires some way of regulating these processes and presumably *coordinating* this regulation. Therefore, the answer to this problem may arise from understanding the organization of signaling pathways from the NMDAR.

*Problem 2:*
*What is the role of molecular events downstream of the NMDAR in the maintenance of the memory trace?*

Is the trace found in the change of synapse strength (synaptic plasticity) or are other cellular processes involved? Can this issue be illuminated by studying the signaling processes downstream of the NMDAR?

To begin to address these questions, we propose a hypothesis – the hebbosome hypothesis – and then describe how it may provide answers to these questions.

## The hebbosome hypothesis

The hypothesis can be stated as follows:

*Synapses contains multiprotein signaling complexes – called hebbosomes – that are responsible for both detecting patterns of neuronal activity and converting this information into a set of cell biological changes altering synaptic and neuronal properties. Hebbosomes function is essential for converting electrical information into the cellular and molecular basis of learning.*

One can paraphrase Hebbs postulate: hebbosomes respond to the patterns of firing where "cell A is near enough to excite a cell B and repeatedly or persistently takes part in firing it." Hebbosomes then specifically mediate and orchestrate the "growth process or metabolic change [that] takes place in one or both cells such that As efficiency, as one of the cell firing B, is increased."

The function of hebbosomes is to translate the information in the neural code into the cellular changes underpinning learning. The organization of signaling proteins into hebbosomes confers functions of signal integration, which facilitate orchestration of multiple downstream cellular processes. Variations in the structure of hebbosomes between different synapses, during development or as part of remodeling of the synapse with activity or pathology, will confer altered cellular responses to the incoming neural code.

## Evidence to support the hebbosome model

In this section we summarize key data that provide the experimental basis for the hebbosome hypothesis.

### Genetic evidence for a signaling complex controlling plasticity and learning

Although it has long been recognized that signal transduction from growth factor receptors and other receptors is mediated by interactions with intracellular proteins (for example, tyrosine kinases; reviewed in Pawson and Scott 1997; Schlessinger 2000) and that ion channels interact with and are modulated by kinases and phosphatases and other signaling proteins (Swope et al. 1999), mouse genetic studies were required to establish that multiprotein channel complexes were important for plasticity and learning.

The first direct evidence that a postsynaptic multiprotein complex could be involved in plasticity and learning emerged from the study of synaptic plasticity in mice carrying a mutation in the Post Synaptic Density 95 (PSD-95) gene (Migaud et al. 1998). PSD-95 is an abundant postsynaptic protein capable of binding many proteins, including the intracellular tail of the NMDAR 2A and 2B subunits (Kornau et al. 1995). Specific deletion of the intracellular tail of NR2 subunits also results in a change in LTP consistent with its function in recruiting a multiprotein complex to the receptor (Sprengel et al. 1998).

Although the synaptic NMDAR currents were unaltered in the PSD-95 mutant mice, there were striking changes in NMDAR-mediated synaptic plasticity, as trains of stimuli ranging from 1-100 Hz induced robust LTP of a greater magnitude than that seen in wild type mice. This plasticity phenotype suggested that PSD-95 was critical in some "negative effector pathway," since its disruption lead to enhancement of LTP. As will be discussed below, this finding opens the possibility that signaling pathways are directly coupled to the NMDAR and their differential recruitment may regulate the direction of plasticity (potentiation or depression).

A second phenotype of the PSD-95 mutant mice was their severe spatial learning deficit (Migaud et al. 1998), a form of learning that is dependent on the NMDAR. These studies of PSD-95 mutants reopened a debate about a fundamental feature of Hebbian models of learning, which in their simplest version argue that activity-dependent strengthening of synapses provides the basis through which synapses can host the memory trace. Despite the relative ease of producing LTP in these mice, the PSD-95 mutants had severe spatial learning impairments. This paradox (discussed in Migaud et al. 1998) will be pursued later, in the context of the relative involvement of the processes downstream from the NMDAR to learning.

The overall conclusion from the PSD-95 mutant mouse study was that the NMDA receptor functions in synaptic plasticity and learning by interacting with PSD-95. More specifically, the NMDAR could be viewed as a component of a signaling complex where the intracellular proteins mediate important signaling functions. This raised the possibility that the complex of NMDAR and PSD-95 contained other important signaling proteins.

## Isolation of 2000kD multiprotein signaling complexes

To address the questions raised about the functions of the protein complex between NMDAR and PSD-95, it was necessary to use biochemical methods that were capable of describing the size and composition of the complexes isolated from intact brain tissue (Husi et al. 2000; Husi and Grant 2001a, b). This strategy would allow the identification of proteins within the complexes and provide some insight into the organization of the NMDA receptor with respect to other receptors and postsynaptic components. Methods were optimized for large-scale isolation of the complexes from mouse brain and proteomic tools were used to identify protein constituents. Several affinity isolation methods aimed at NMDAR subunits and

PSD-95 isolated similar complexes, which were physically distinct from other synaptic complexes, such as AMPA receptor complexes.

The NMDA receptor complexes (NRC) were found to range in size between 2000 and 3000 kDa, which is a similar size to a ribosome. This mass is several times greater than would be expected for just the NMDAR channel subunits, and protein staining indicated that the complexity of the NRC was high. This finding was confirmed in subsequent proteomic analysis that revealed at least 75 proteins in NRCs (details of this composition and discussion can be found in Husi et al. 2000). During the course of this biochemical study, a subset of the proteins identified using the proteomic approach was found using the yeast two-hybrid system (Husi et al. 2000). We consider that the NRC described in this way is a prototypical hebbosome and that other receptor complexes will have similar features.

### Proteomic evidence for a signaling complex controlling plasticity and learning

The surprising diversity of proteins uncovered by the proteomic study has lead to the model that the 2000-kDa complex of proteins is a signaling device responsible for a number of features of plasticity and learning.

### NMDA receptor complexes contain five classes of protein

The proteins could be categorized into five general classes:
1) neurotransmitter receptors,
2) adaptor proteins,
3) signaling proteins,
4) cell adhesion proteins and
5) cytoskeletal proteins.

These broad categories of proteins give a general picture of a subsynaptic structure that serves a general function, in the same way one considers other protein complexes in cell biology such as transcription and splicing complexes. Each of the five classes deserves brief discussion.

### 1. Neurotransmitter receptors

The NMDAR and mGlu receptors were found in the complexes, in contrast to the AMPA receptors, which were absent. This finding immediately provides an insight into the function of the complexes in synaptic plasticity, since it is well known that NMDARs and mGluRs regulate the induction of plasticity, in contrast to the AMPA receptor, which is involved in mediating the fast EPSP and the expression of plasticity. The notion that both multiple, distinct neurotransmitter receptors could be components of one complex sets up the possibility that the signaling of these two classes (ionotropic and metabotropic) of receptors could be integrated within the complex.

## 2. Adaptor proteins

In addition to PSD-95 and the related membrane associated guanylate kinase (MAGUK) proteins, several other classes of adaptor were found (GKAP, SHANK, Yotiao, AKAPs, APPL). Although these proteins are known to mediate linkages between other proteins (Naisbitt et al. 1999), it should be noted that many of the proteins in the complexes have multiple protein-protein interaction domains and can perform adaptor or scaffold functions. Bioinformatic studies (H. Husi and S. Grant, unpublished) indicate a rich array of potential protein interactions between proteins in the complexes, including those mediated by adaptors or scaffold proteins.

## 3. Signaling proteins

The surprising diversity of signaling proteins has lead to several themes that together present the complexes as an enzymatically active machine. First, calcium-sensitive enzymes that could respond to calcium from the NMDAR in a "microdomain" for sensing the pulses of calcium influx were found. Second, some general cellular signaling enzymes (e.g., protein kinase A and MAP kinase) were detected, consistent with the view that their recruitment to this complex may subserve some very local function. This function could be in responding to the receptor activation in the complex or in phosphorylating substrates within the complex. Third, and perhaps most interesting, was that sets of signaling proteins that are known to comprise pathways (e.g., Ras-Raf-MEK-MAPK) were found. This finding raises the possibility that signal processing in the complex may occur at the level of pathways. Patterns of firing detected by the receptors may recruit one or other pathway, which then may control some specific feature of plasticity. Moreover, the presence of multiple pathways could allow cross talk between pathways within the complexes.

## 4. Cell adhesion proteins

The presence of adhesion proteins that are known to mediate homophilic cell-cell interactions and signaling may indicate that the complexes are tethered in specific locations at the postsynaptic side of the synapse with respect to the presynaptic terminal. This may be important for aligning the presynaptic release machinery with the postsynaptic receptor signaling complex, for providing some alternative mechanism for signaling between neurons, and for stabilizing synapses during mechanical events such as shape remodeling.

## 5. Cytoskeletal proteins

Several cytoskeletal proteins, some of which interact directly with NMDAR subunits, were present in the complexes. These proteins most likely regulate the assembly of the complexes, and some aspects of their dynamic features, with respect to the other synaptic and neuronal architecture.

## Physiological evidence for functions of the five classes of complex proteins in synaptic plasticity

What is the general function of the NRC? This question can be addressed by examining the role of its individual components. The first observation is that as many as 30 proteins in the NRCs are involved with the induction of synaptic plasticity. Table 2 shows the five categories of proteins and refers to genetic and pharmacological data indicating evidence of their role in synaptic plasticity. Among these 30 proteins are many of the signaling proteins that have garnered the most attention over the last 15 years: protein kinases (PKA, PKC, CamkII, tyrosine kinases, ERK) and phosphatases (PP1, PP2B) and mGluR. This finding was surprising, given that these enzymes have been shown to be involved with the induction step (converting the activity into the cellular events in the first 30 minutes after activating the NMDAR). The second observation was that proteins

**Table 2.** NRC proteins involved with synaptic plasticity. The five categories of NRC proteins (Husi et al. 2000; Husi and Grant 2001b) are listed in the left column. Specific proteins in these categories are listed in the other columns where they are known to be involved with the induction of synaptic plasticity. The three columns distinguish three types of interference experiments; targeted gene disruption using homologous recombination in mouse embryonic stem cells (Targeted mutation), transgenic expression using randomly integrated transgenic constructs in mice (Transgenic expression), and pharmacological interference using drugs specific for the named proteins (Pharmacological interference).

|  | Targeted Mutation | Transgenic Expression | Pharmacological Interference |
|---|---|---|---|
| 1. Neurotransmitter Receptor | NR1 NR2A mGluR1 | NR2B | NR mGluR |
| 2. Cell adhesion |  | L1 . | L1 N-Cadherin |
| 3. Adaptors | PSD-95 |  |  |
| 4. Signalling | PKA catalytic PKA regulatory PKC CamKII nNos Ras NF-1 SynGAP | CamKIIα Calcineurin PKA inhibitor Ras Rap | CamKII PKA PKC Tyrosine Kinase MEK PI3-kinase PP1 PP2A PP2B Calmodulin NOS cPLA2 |
| 5. Cytoskeleton |  |  | Actin polym. |

from each of the five categories of the NRC sets have been shown to be important for plasticity. In other words, disruption of any of the five general molecular components of the complexes results in impairments in the overall function of the NRC – the induction of plasticity.

This analysis provides a compelling case that the complexes are a *machine* or *device* that can be interfered with in many different ways. This suggests that the components of the hebbosome complex serve a common function, which is the *emergent* property of the complex itself rather than of the individual components. This finding has lead us to move away from a widely used concept in plasticity and learning; the concept of the "key molecule" or "molecular switch." Any one of the individual components of hebbosomes could be considered "key molecules" if their disruption alters the overall function of the complex (perhaps in the same way that most parts of an automobile when broken may impair the overall ability of the car to drive across town).

**The NRC and the induction step of LTP.** The subdivision of the "induction of LTP into the two underlying functions of "pattern detection" and "instruction of change" is helpful in understanding the relationship of the NMDAR itself to the other proteins in the signaling complex. The NMDAR subunits are clearly capable of distinguishing different degrees of postsynaptic activity by virtue of their voltage-dependent Mg2+ regulation. High frequency trains, which are associated with greater postsynaptic depolarization than low frequency trains, are more likely to produce LTP (Herron et al. 1986), since a greater degree of $Ca^{2+}$ influx occurs. The NR2 subunit composition of NMDA receptors is also important for determining the $Ca^{2+}$ influx, and interference with specific NR2 receptor subunits can alter LTP induction Seeburg et al. 2001). Subunit composition has been correlated with developmental changes in visual and hippocampus plasticity (Philpot et al. 2001; Seeburg et al. 2001).

The fact that PSD-95 and other complex proteins do not have any effect on synaptic NMDA currents, yet have clear effects on the induction of plasticity, indicates that the degree of NMDAR activation alone is not responsible for the induction of plasticity. Moreover, it implies that the degree of activation of associated proteins determines the induction of plasticity. The simplest way to reconcile these data is by postulating that the NMDAR is involved with pattern detection (mostly by its ability to gate $Ca^{2+}$ influx) and that the plasticity pathways recruited by $Ca^{2+}$ influx are within the associated protein complex. Under normal physiological conditions, the combination of $Ca^{2+}$ influx and the degree of pathway activation sums to produce the output of the induction step.

The combination of pattern detection by the NMDAR and instruction of intracellular signaling by the associated signaling proteins also provides two general levels of modulation for the induction step. The modulation of the NMDAR subunits by phosphorylation (e.g., by PKA, PKC, Src and other kinases; PP1, PP2A and other phosphatases) is recognized as a means of altering $Ca^{2+}$ influx. Although less well characterized, phosphorylation of many of the associated proteins, which are important in plasticity, also occurs. Some of these kinases, for example Fyn, can phosphorylate both NMDAR and MAGUK proteins, and thus the modulatory effects of phosphorylation may be distributed across several points of regulation within the complexes. In so doing, the kinase can modulate both the $Ca^{2+}$ trigger event via the channel and the downstream pathway coupling this signal to the relevant effector mechanisms.

The ability of synapses to undergo a variety of forms of NMDAR-dependent plasticity – bi-directional plasticity and metaplasticity – has attracted attention, particularly in the context of developmental plasticity, where low frequency trains are capable of inducing LTD (Heynen et al. 2000; Abraham et al. 2001). Several lines of data indicate that this ability may in part result from conditions of lower $Ca^{2+}$ influx relative to those conditions that induce LTP. In this model, there is no need to have anything more than a single effector pathway downstream of the NMDAR, and it is the relative degree of its activity that could determine a threshold between LTD and LTP. An alternative view is that the difference between LTP and LTD induction is regulated by the extent of activation of *distinct* pathways – positive and negative effectors within the complex.

This view is supported by several lines of evidence. First, interference with some proteins in the complex has a more marked effect on LTP induction than on LTD induction, and vice versa. Second, the mutations in PSD-95 and other complex proteins indicate that there are separate pathways downstream of the NMDAR (Migaud et al. 1998) and that uncoupling of one pathway can radically alter the frequency function for LTP induction. For example, PSD-95 mutations result in enhanced LTP when 1Hz trains are applied, and this has no effect on wild type slices (Migaud et al. 1998).

The activation of the NMDAR by trains of stimuli does not necessarily lead to change in synaptic strength, particularly when low frequency or brief activation is involved. On one hand, it could be argued that the lack of change in synaptic strength indicates these patterns of stimuli are not being detected by the NMDAR and associated proteins and therefore there is no change in synaptic properties. On the other, it may be that, under these conditions, a negative "restraining" mechanism is actually activated and prevents LTP induction. This phenomenon has been observed and described as an example of "metaplasticity" (the plasticity of synaptic plasticity), where brief activation of the NMDAR leads to a transient inhibition of LTP for 1 hour, after which LTP induction returns to normal (Huang et al. 1992; Abraham and Bear 1996). Moreover, the phenotype of the PSD-95 mutants appears to underpin this type of mechanism since LTP was induced by 1 Hz trains that normally do not induce LTP (Migaud et al. 1998). Together these findings suggest that the complex is monitoring patterns of activity and distinguishing those that are capable of inducing LTP from those that have other modulatory properties.

### Signal integration and pathways within the complex

The ability of the complex to respond differentially to distinct patterns of activity raises the exciting possibility that the complex contains machinery for computing the salience of a signal and triaging this signal to the downstream effector mechanisms. The diversity of signaling proteins within the complex may provide a basis for the coupling of distinct pathways to the neurotransmitter receptors.

Key to this aspect of the hypothesis is to demonstrate that distinct plasticity signaling pathways within Hebbosomes are coupled to the NMDAR. This point is illustrated by mutations in PSD-95 and its binding protein, SynGAP, and their coupling to the MAPK pathway (Migaud et al. 1998; Komiyama et al. 2002). Within the complex are the components of the Ras-MAPK pathway (Ras, Raf, MEK, MAPK); this pathway is regulated by SynGAP (a GTPase activity protein). Although SynGAP couples the MAPK pathway to the NMDAR via PSD-95 and is required for LTP, PSD-95 also couples another MAPK-independent pathway. Thus PSD-95 is at a point of bifurcation of downstream pathways. Moreover, this bifurcation also controls the bi-directional features of synaptic plasticity, since PSD-95 mutations lead to enhanced LTP and mutations in SynGAP lead to reduced LTP. In addition to illustrating how multiple pathways are within the complexes, this finding also indicates that intra-complex cross talk of pathways could allow computation of a diverse set of signaling outcomes. As mentioned earlier, the complex can receive signals from both NMDA and mGlu receptors that further add to this model by suggesting multiple input signals. These data lead us to propose that the complex has multiple inputs and multiple internal pathways and output pathways, and the role of the complex is to process these signals and orchestrate the response in the neuron (Fig. 1).

### Outputs from the complex

The cell biological events or mechanisms that respond to signaling from the NMDA receptor are diverse in type and subcellular localization (Problem #1, above). At the level of the postsynaptic terminal there is trafficking of AMPA receptors, protein translation and cytoskeletal changes. More broadly, there are changes in the presynaptic terminal and neuron (mediated via retrograde signals), in local dendritic events, and in global gene expression in the nucleus These observations lead us to ask if the complexes are regulating these events and could this regulation be achieved by specific proteins and pathways organized by the complexes? Although this has yet to be shown using genetic techniques, it is already clear that distinct pathways are regulated by the interacting proteins.

Although the NMDAR is known to regulate these downstream effectors, and that by extension of this argument it could be suggested that the NRC is important, further support of this hypothesis is given by examination of the specific NRC proteins. Specific enzymes within these complexes are known regulators of translation, actin cytoskeleton, AMPA trafficking and transcription. This fact suggests that these enzymes in the NRC can drive these other downstream processes, either by contact with the enzymes in the NRC or translocation of protein components. A potentially important physiological aspect of the multiple pathways from the NRC orchestrating these downstream mechanisms is the idea that different patterns of synaptic activity (e.g., 1 Hz or 5 Hz trains) might lead to differential recruitment of these mechanisms. This process would allow the NRC to link some patterns of synaptic activity to translation and others to, say, gene expression and AMPA trafficking.

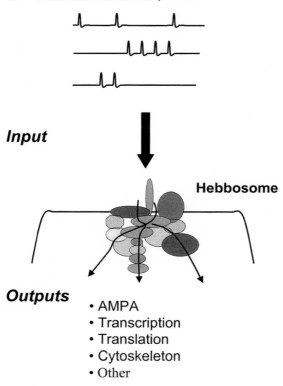

**Input**

**Hebbosome**

**Outputs**
- AMPA
- Transcription
- Translation
- Cytoskeleton
- Other

**Fig. 1.** Hebbosome signaling properties. Patterns of action potentials are illustrated as the traces in the top part of the figure. These act as the input to the hebbosome, which is shown as a complex of proteins on the postsynaptic side of the synapse. Multiple neurotransmitter receptors may be components of the hebbosome and thus act as multiple specific inputs (e.g., metabotropic and NMDAR subunits in the one complex). Three arrows within the hebbosome indicate that multiple biochemical pathways initiated by activation of the receptors are driving distinct downstream events as well as participating in cross talk. In other words, signals are integrated or orchestrated within the hebbosome. These pathways act as outputs to downstream effector mechanisms, of which several examples are mentioned (AMPA receptor phosphorylation and trafficking, transcription, etc. These are the same mechanisms listed in Table 1).

### Behavioral evidence for functions of the five classes of complex proteins in rodent learning

As predicted by Hebb in his neurophysiological postulate, there could be some "metabolic or structural" mechanism that might allow the information contained within repeated patterns of firing to lead to strengthening of signaling between two neurons (Hebb 1949). As described above, the complexes appear to perform this function at the synaptic level. To summarize the relevance of the function of the complex to learning, Table 3 shows a list of proteins in the complexes that are involved in rodent learning through pharmacological and genetic experiments. As

**Table 3.** NRC proteins involved with learning in rodents and humans. The five categories of NRC proteins are listed in the left column and the other two columns list proteins involved with learning in rodents and humans. Both genetic and pharmacological interference experiments are combined in the rodent studies, and the proteins listed in humans are heritable mutations. (For further details see Husi et al. 2000; Husi and Grant 2001b)

| Hebbosome Protein | Rodents | Humans |
|---|---|---|
| 1. Neurotransmitter Receptor | NR1<br>NR2A<br>NR2B<br>MGluR1 | |
| 2. Cell adhesion | L1 | L1 |
| 3. Adaptors | PSD-95 | |
| 4. Signalling | PKA<br>PKC<br>αCamKII<br>PP1<br>PP2A<br>PP2B<br>nNOS<br>NF-1<br>MEK<br>SynGAP<br>cPLA2 | NF1<br>RSK2 |
| 5. Cytoskeleton | | |

many as 18 different proteins spread across the multiple categories of proteins are required for normal learning. As is the case for the NMDAR, many of these proteins are involved with the acquisition, rather than the retention or recall, of hippocampus-dependent forms of learning.

These data support the model that this complex, rather than the individual components, is a key device required during learning. The function of the complex in detecting and acting upon specific patterns of activity and then driving changes in cell biology would fit the need for a learning model that uses activity-dependent changes at the synapse. The large number of molecules involved in both the *induction* of plasticity and learning indicates that the same complex is involved in both processes.

### Evidence for functions of the complexes in human learning

Given the enormous body of literature on NMDAR function in model organisms, there is very little information on its role in human learning. The proteomic study of the NRC has built a new bridge to the human biology since three NRC proteins were previously known to underlie human heritable learning disability (Husi et al. 2000). These proteins were not in the NMDA channel but in the downstream signaling proteins RSK-2 and NF1 and in the adhesion protein, protein L1. This location suggests that the associated proteins may be particularly important in human disorders and are excellent candidate genes for human genetic studies. Studies of these genes in humans with psychiatric disorders are underway and may lead to the identification of new diagnostics and understanding. Understanding the intricacies of the signaling networks within hebbosomes could provide new therapies in which, for example, the deleterious effects of a hebbosome mutation or drug interference are ameliorated by pharmacological targeting of some other hebbosome protein.

## A model for the function of the complexes in plasticity and learning

The following is a description or synthesis of the evidence described above, illustrating how the complexes are imagined to work under simple conditions of a two-neuron system.

Neuron A is presynaptic to B and during learning A is repeatedly firing and driving activity in B. Before the salient learning event occurs, cell A fires occasionally (at low frequency), and the action potential invades the presynaptic terminal of cell A and releases glutamate onto the postsynaptic terminal on cell B. This release is detected by the postsynaptic glutamate receptors; however, it is not sufficient for triggering the biochemical pathways in the complexes that will lead to changes in synaptic strength.

As a salient learning event occurs and induces a different (e.g., high frequency) pattern of firing in cell A, this causes a different pattern of activation of the complexes in cell B, where the complex is triggered by the influx of calcium via the NMDAR and mGluR activation events. Under these conditions, multiple enzymes in the complexes are activated, which then initiate signaling cascades, which ultimately lead to outputs from the complex to downstream cellular effector mechanisms. Importantly, distinct patterns of firing can activate distinct pathways within the complexes. Cross talk between pathways and relative degrees of activation of positive and negative effector pathways are the initial events in coordinating the overall effects of the patterns of activity. These steps are the induction of plasticity.

The hebbosome, as the induction machinery, now connects too and orchestrates the expression of plasticity. The ability of the complex to coordinate multiple types of downstream cellular events is intrinsic to the diversity of proteins (and pathways) within the complexes. In other words, some pathways are well suited to regulate AMPA receptor function, others for cytoskeletal changes and cell adhesion (for structure), others local translation, transcription and so on. The sum total of the effects on these distinct processes is the "expression" of plasticity.

When these induction processes are interfered with, in many different ways, the overall effect on expression is influenced. As described above and in Tables 2 and 3, interference with any of the five categories of molecules in the complexes has an overall impact on the induction of plasticity and acquisition of learning. These observations underpin the notion that the reason for having this complex is to carefully orchestrate the multitude of processes underlying the changes in cell biology resulting from distinct patterns of neural activity.

## Implications and speculations based on the model

This model places emphasis on the notion that a synapse contains machines dedicated to interpreting and acting upon the pattern of neural activity. In so doing, these machines act as input devices to the neuron. The aspect of multiple signal outputs from the complex has implications for the effects of induction on the synapse, the surrounding dendrite and the whole neuron. Combining the knowledge of the multiple pathways in the complex with information showing NMDAR can drive synaptic changes as well as non-synaptic changes suggests that there may be important differences in the biological significance of the distinct sites or locus of change of expression.

The prevailing synaptic plasticity model of learning is one in which the locus of change – the memory trace – is in the resultant strength of the synapses [the synaptic expression of plasticity (SEP) model; see Fig. 2, panel A]. Our model would not conflict with this model in that the complexes are involved with regulating the changes in synaptic strength. Although the details of how AMPA receptors are regulated are still emerging, it is abundantly clear that their function is downstream of the NRC. Thus a situation in which the induction machinery is perturbed would be expected to result in inappropriate changes in AMPA receptors in response to patterns of activity. However, this is not the only way our model could be involved with controlling the locus of memories, and alternative models arise.

To illustrate an extreme alternative, one could imagine that the salient patterns of firing activate the complexes, which then activate several signaling pathways to drive outputs. If we imagine that these outputs do not change the AMPA receptor function (or other functions that result in a change in the strength of this individual synapse), but change some other features of the dendrite or neuron, then maybe these other functions could alter the properties of the circuit and learning. What could these other expression events be? They could be changes in excitability of the presynaptic neuron (Fitzsimonds et al. 1997), changes in the local dendrite (excitability or structure), changes in dendritic mRNA or protein trafficking, or global changes in gene expression.

This model, which we refer to as the distributed expression of plasticity (DEP) model (Fig. 2, panel B), would also be consistent with the data linking interference with the complexes with impaired learning. The DEP model may appear to be disadvantageous from the point of view of losing information gained by storing information in individual synapses; however, it is not entirely clear if absolute synapse specificity is either required for learning or indeed occurs with learning. The NRC complexes may be well placed to induce DEP changes over local areas of dendrite and may include adjacent synapses. Some components

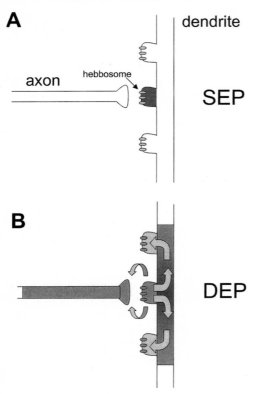

**Fig. 2.** Hebbosomes and the DEP model. The model of hebbosomes regulating multiple downstream pathways raises the question as to the significance of those pathways for the neuron in learning. Panel A. The synaptic expression of plasticity (SEP) model is the traditional model of learning in which the change in synaptic strength is the locus of learning. This model has been updated to included hebbosomes. Patterns of action potentials arrive at the synapse and activate postsynaptic hebbosomes (illustrated as three small structures on the dendritic spine). The result of their signaling activity is to lead to changes in the strength of the specific synapse in which they reside. This is illustrated as the shading of this dendritic spine and not the adjacent spines. Panel B. In contrast to the SEP model, the distributed expression of plasticity (DEP) model has changes in cellular function distributed to regions other than the specific synapse that received the inputs. Patterns of action potentials arrive at the synapse and activate hebbosomes, which then send multiple downstream pathways (indicated by arrows; see also Fig. 1) to change the properties of the synapse, local dendrite, neighboring synapses, presynaptic terminal and cell nucleus. These effects (see Table 1) are orchestrated by hebbosome signaling properties.

may regulate cytoskeleton and spine number. In this way, graded changes over a region of dendrite may readily arise, which could be desirable from the point of view of establishing cognitive maps. Thus some patterns of activity could induce synapse-specific changes and perhaps others could produce a spread of changes in local synapses. Another reason why maintenance of changes solely at the synaptic level may be undesirable comes from recent observations on the extraordinarily dynamic nature of synaptic proteins as well as the rapid shape changes in the synapse itself. Indeed, the AMPA receptors, which are the best candidates for maintaining the level of synaptic strength, are rapidly turned over and regulated by several biochemical mechanisms involving transient phosphorylation and protein-protein interactions. It is possible that the dynamic changes could undermine the ability to store information over periods of years. Although we consider it possible that the DEP model is correct under some circumstances, the simultaneous expression of synaptic plasticity could allow for "homeostatic" regulation of synaptic strength.

## Conclusions

We have attempted to focus attention on the role of multiprotein signaling complexes that contain the fundamental functions of action potential pattern detectors and master regulators of cell state. The available information now allows the hypothesis that these functions can be contained within a single isolatable structure, the hebbosome. As already indicated, hebbosomes are not a singular entity but reflect a family of complexes in which diversity in structure and location infer subtly different properties in various regions of the nervous system. Indeed, other neurotransmitter receptor complexes are now being identified with similar molecular complexity to the NRC (Kim et al. 2001). This hebbosome hypothesis largely explains many of the features of hippocampus synaptic plasticity and its relationship to learning. The broader implications of this hypothesis is to identify a molecular machine underlying a fundamental feature of the nervous system and to now explore its features in human cognition and psychiatric disease.

The notion that hebbosomes are monitoring the patterns of synaptic activity and triaging intracellular signals to modify various features of the synapse and neuron underlines their evolved function in encoding learning and memory. With this model it is not surprising that interference in either of the two key functions (pattern detection and instruction of cellular changes) would result in learning impairments in the intact animal (see Table 3). Thus the hebbosome can be thought of as a device for plasticity and learning. Despite the unification of some plasticity and learning functions within the structure of the hebbosome, it is still premature to conclude that the change in synaptic strength is necessarily the substrate of learning, as opposed to some other type of neuronal plasticity controlled by hebbosomes.

## Acknowledgments

This work was supported by the Wellcome Trust, European Community, Pew Charitable Trusts and National Institute of Mental Health. We thank the members of our laboratories for the many conversations on the subject of signaling complexes.

# References

Abraham WC, Bear MF (1996) Metaplasticity: the plasticity of synaptic plasticity. Trends Neurosci 19:126–130

Abraham WC, Mason-Parker SE, Bear MF, Webb S, Tate WP (2001) Heterosynaptic metaplasticity in the hippocampus in vivo: a BCM-like modifiable threshold for LTP. Proc Natl Acad Sci USA 98:10924–10929

Adrian (1928) The basis of sensation, the action of the sense organs. New York, W. W. Norton & Company Inc.

Bliss TV, Lomo T (1973) Long-lasting potentiation of synaptic transmission in the dentate area of the anaesthetized rabbit following stimulation of the perforant path. J Physiol 232:331–356

Bliss TV, Collingridge GL (1993) A synaptic model of memory: long-term potentiation in the hippocampus. Nature 361:31–39

Collingridge GL, Kehl SJ, McLennan H (1983) Excitatory amino acids in synaptic transmission in the Schaffer collateral-commissural pathway of the rat hippocampus. J Physiol (Lond) 334:33–46

Finger S (1994) Origins of neuroscience: a history of explorations into brain function. New York: Oxford University Press

Fitzsimonds RM, Song HJ, Poo MM (1997) Propagation of activity-dependent synaptic depression in simple neural networks. Nature 388:439–448

Grant SG, Karl KA, Kiebler MA, Kandel ER (1995) Focal adhesion kinase in the brain: novel subcellular localization and specific regulation by Fyn tyrosine kinase in mutant mice. Genes Dev 9:1909-1921

Grant SG, ODell TJ, Karl KA, Stein PL, Soriano P, Kandel ER (1995) Impaired long-term potentiation, spatial learning, and hippocampal developement in fyn mutant mice. Science 262:760–763

Hebb DO (1949) The organization of behavior; a neuropsychological theory. New York, Wiley

Herron CE, Lester RA, Coan EJ, Collingridge GL (1986) Frequency-dependent involvement of NMDA receptors in the hippocampus: a novel synaptic mechanism. Nature 322:265–268

Heynen AJ, Quinlan EM, Bae DC, Bear MF (2000) Bidirectional, activity-dependent regulation of glutamate receptors in the adult hippocampus in vivo. Neuron 28:527–536

Huang YY, Colino A, Selig DK, Malenka RC (1992) The influence of prior synaptic activity on the induction of long-term potentiation. Science 255:730–733

Husi H, Grant SG (2001a) Isolation of 2000-kDa complexes of N-methyl-D-aspartate receptor and postsynaptic density 95 from mouse brain. J Neurochem 77:281–291

Husi H, Grant SG (2001b) Proteomics of the nervous system. Trends Neurosci 24:259–266

Husi H, Ward MA, Choudhary JS, Blackstock WP, Grant SG (2000) Proteomic analysis of NMDA receptor-adhesion protein signaling complexes [see comments]. Nature Neurosci 3:661–669

Kandel ER (2001) The molecular biology of memory storage: a dialogue between genes and synapses. Science 294:1030–1038

Kim M, Jiang LH, Wilson HL, North RA, Surprenant A (2001) Proteomic and functional evidence for a P2X7 receptor signalling complex. Embo J 20:6347–6358

Komiyama NH, Watabe AM, Carlisle HJ, Porter K, Monti J, Strathdee DC, O'Carroll CM, Martin SJ, Morris RGM, O'Dell TJ, Grant SGN (2002) SynGAP regulates MAPK signalling, synaptic plasticity and learning in a multiprotein complex with PSD-95 and NMDA receptor. Journal of Neuroscience 22(22): 9721–9732

Kornau HC, Schenker LT, Kennedy MB, Seeburg PH (1995) Domain interaction between NMDA receptor subunits and the postsynaptic density protein PSD-95. Science 269:1737–1740

Malenka RC, Kauer JA, Perkel DJ, Mauk MD, Kelly PT, Nicoll RA, Waxham MN (1989) An essential role for postsynaptic calmodulin and protein kinase activity in long-term potentiation. Nature 340:554–557

Malinow R, Madison DV, Tsien RW (1988) Persistent protein kinase activity underlying long-term potentiation. Nature 335:820–824

Migaud M, Charlesworth P, Dempster M, Webster LC, Watabe AM, Makhinson M, He Y, Ramsay MF, Morris RG, Morrison JH, O'Dell TJ, Grant SG (1998) Enhanced long-term potentiation and impaired learning in mice with mutant postsynaptic density-95 protein [see comments]. Nature 396:433–439

Naisbitt S, Kim E, Tu JC, Xiao B, Sala C, Valtschanoff J, Weinberg RJ, Worley PF, Sheng M (1999) Shank, a novel family of postsynaptic density proteins that binds to the NMDA receptor/PSD-95/GKAP complex and cortactin. Neuron 23:569–582

O'Dell TJ, Kandel ER, Grant SG (1991) Long-term potentiation in the hippocampus is blocked by tyrosine kinase inhibitors. Nature 353:558–560

Pawson T, Scott JD (1997) Signaling through scaffold, anchoring, and adaptor proteins. Science 278:2075–2080

Philpot BD, Sekhar AK, Shouval HZ, Bear MF (2001) Visual experience and deprivation bidirectionally modify the composition and function of NMDA receptors in visual cortex. Neuron 29:157–169

Sanes JR, Lichtman JW (1999) Can molecules explain long-term potentiation? Nature Neurosci 2:597–604

Schlessinger J (2000) Cell signaling by receptor tyrosine kinases. Cell 103:211–225.

Seeburg PH, Single F, Kuner T, Higuchi M, Sprengel R (2001) Genetic manipulation of key determinants of ion flow in glutamate receptor channels in the mouse. Brain Res 907:233–243

Sherrington CS (1906) The integrative action of the nervous system. New York, C. Scribner's Sons

Silva AJ, Stevens CF, Tonegawa S, Wang Y (1992) Deficient hippocampal long-term potentiation in alpha-calcium-calmodulin kinase II mutant mice. Science 257:201–206

Sprengel R, Suchanek B, Amico C, Brusa R, Burnashev N, Rozov A, Hvalby O, Jensen V, Paulsen O, Andersen P, Kim JJ, Thompson RF, Sun W, Webster LC, Grant SG, Eilers J, Konnerth A, Li J, McNamara JO, Seeburg PH (1998) Importance of the intracellular domain of NR2 subunits for NMDA receptor function in vivo. Cell 92:279–289

Swope SL, Moss SI, Raymond LA, Huganir RL (1999) Regulation of ligand-gated ion channels by protein phosphorylation. Adv Second Messenger Phosphoprotein Res 33:49–78

Waddell S, Quinn WG (2001) Flies, genes, and learning. Annu Rev Neurosci 24:1283–1309

# Analysis of Brain Disorders Using DNA Microarrays

*K. Mirnics*[1,2,4*], *F.A. Middleton*[2], *J.N. Pierri*[1], *D. A. Lewis*[1,3] *and Pat Levitt*[2,4]

## Summary

Recent advances in DNA microarray technology have enabled researchers to simultaneously assess complex gene expression changes for tens of thousands of genes, making this an optimal approach for studying brain diseases. Discovered transcriptome changes are likely to reveal new information about the pathophysiology of disease states and identify novel genes that suggest previously unknown molecular interactions. In the years ahead, gene expression profiling may allow for the individualization of diagnosis, ultimately linking molecular phenotypes to the specific clinical profiles. Defining disease-specific transcriptome changes will also facilitate development of animal models that closely mimic common molecular events associated with schizophrenia, depression, autism or other mental disorders. In turn, this approach will lead to the improved application of pharmacogenomics, where compounds are screened for their ability to modulate the transcripts that are changed in a diagnosis-specific manner.

In our recent studies using cDNA microarrays, we uncovered three types of altered gene expression patterns in the prefrontal cortex of individuals with schizophrenia when compared to matched controls. First, functional data mining led to the discovery of consistent decreases in several functional groups consisting of transcripts encoding proteins that regulate presynaptic function, neurotransmitter signaling and a restricted number of metabolic pathways. Second, we identified a number of known genes that showed consistently changed expression in schizophrenia and that had not been previously associated with schizophrenia.

Finally, we also identified altered expression of several genes with no known function but with distinct brain expression patterns. Functional studies of these molecules are in progress. The combination of findings reveals the strengths of the gene microarray strategy to guide future studies for deciphering complex molecular defects in mental disorders.

Departments of [1]Psychiatry, [2]Neurobiology, [3]Neuroscience and [4]PittArray, University of Pittsburgh School of Medicine, Pittsburgh, PA 15261
*Correspondence:* Karoly Mirnics, Dept. of Psychiatry, U. of Pittsburgh School of Medicine, W1655 BST, Pittsburgh, PA 15261, USA;
phone: 412-648-9788; fax: 412-648-1441; e-mail: karoly+@pitt.edu

Mallet/Christen
Neurosciencess at the Postgenomic Era
© Springer-Verlag Berlin Heidelberg 2003

## The Genomic Revolution

In the last decade, the scientific community finished the draft of the human genome (Lander et al. 2001; Olivier et al. 2001) and we entered the post-genomic era(Husi and Grant 2001), initiating the exploration of "functional genomics" and "proteomics" (Stoll et al 2002; Dongre et al. 2001). Students learn about "transcriptome profiling" and "conditional genetic ablations" (Carninci et al. 2001), while gene cloning and transfection have become routine procedures in many neuroscience laboratories. Northern hybridizations are giving way to microarray studies; PCR has evolved into a real-time quantitative assay (Freeman et al. 1999; Brail et al. 1999), while mass-sequencing of nucleic acids has led to development of serial analysis of gene expression (SAGE; Velculescu et al. 1995). These high throughput methods are rapidly changing the way we plan and perform experiments. We are living in an unprecedented scientific revolution, where the most optimistic deadlines turn out to be too conservative: the Human Genome Project produced a draft of the whole human genome ahead of time (Lander et al. 2001; Husi and Grant 2001); most of the single gene diseases have been identified, and gene therapy holds the promise of creating healthy babies who would have died from inherited disorders (Pfeifer and Verma 2001; Barranger and Novelli 2001).

## The Microarray Technology

DNA microarrays are a crucial part of this technical revolution (Schena et al. 1995; DeRisi et al. 1996; Lockhart et al. 1996). Microarrays are based on the well-established principle of complementary hybridization between nucleic acids (Southern et al. 1999), taking advantage of a miniaturization process and development of high-throughput printing robots. A fragment of DNA is synthesized onto glass or deposited on it, and represents the microarray "probe". The "target" is a labeled sample, and the probe will bind to the target if there is a large degree of complementarity between them. The target is labeled by using either radioactivity or fluorescent markers, and its emitted signal can be assessed using appropriate methods. The probe is always in excess so that, the more target it binds, the stronger the emitted signal from the probe-target complex becomes. As such, the signal intensity is proportional to the abundance of the labeled target gene in the sample.

A typical, high-density gene expression microarray contains many thousands of microarray probes, simultaneously assessing information from a significant portion of the whole genome (Mirnics et al. 2001a,b; Marcotte et al. 2001; Luo and Geschwind 2001). Microarray probes may be constructed using shorter oligonucleotides (25–75; Lockhart et al. 1996; Lockhart and Barlow 2001a,b) or longer pieces of DNA (up to several kB) that can be generated using a variety of procedures (Mirnics et al. 2001 a; Luo and Geschwind 2001). Subtractive hybridization, representational difference analysis (Geschwind et al. 2001) and tissue-specific library clones have been used successfully as microarray probes (Alizadeh et al. 2000). Oligonucleotide arrays usually analyze each target sample on a single microarray and the comparison between different microarrays (samples) is performed post hoc using mathematical algorithms (Lockhart and Barlow 2001a; Lipshutz et al. 1999). In contrast, cDNA arrays are run most often as dual-fluorescent hybridizations, where the two samples to be compared (or a sample and reference) are labeled with different fluorescent dyes and hybridized onto the same microarray

(Marcotte et al. 2001; Luo and Geschwind 2001). This procedure allows a direct comparison between the two samples and, for certain experimental designs, this approach may be advantageous (Mirnics 2001, 2002). Oligonucleotide arrays and cDNA arrays both perform well and generate excellent datasets, yet both share a number of limitations (Mirnics et al. 2001a,b; Mirnics 2002). As an alternative, membrane based cDNA "macro" arrays remain a viable choice for analysis of gene expression (Bertucci et al. 1999; Becker 2001; Vawter et al. 2001).

## Gene Expresion Microarrays in Brain Studies

The first successful microarray experiments were performed in the mid-1990s (Schena et al. 1995; Lockhart et al. 1996) and triggered a flurry of cancer-related microarray studies. Yet, reports using brain tissue in microarray studies were absent for years after the initial publications. Because of the need to use postmortem tissues for many human brain diseases, obtaining high quality, well-identified sample tissue was a significant limitation. The phenotypic complexity of brain tissue also presented a major hurdle in transcript detection and data analysis. Furthermore, in most brain disease processes, only a fraction of cells is affected and the transcript changes are "masked" by the transcripts originating from unaffected cells. Finally, gene expression changes associated with brain diseases are likely to be moderate in magnitude and often do not exceed a twofold change, even though a 30% expression change may have a profound physiological effect on brain function. Thus, many biologically relevant expression changes would probably remain unnoticed in microarray experiments.

Indeed, the first studies involving microarrays and brain tissue illustrated both the promise and the limitations of this technology. In a microarray study involving animals whose tissues were harvested during sleep and wake cycles, Cirelli and Tononi (1999) observed only a few expression changes. Similarly, in a landmark study, Sandberg et al. (2000) discovered previously unknown expression differences across brain regions – yet, the low number of reported differences belied previously reported differential gene expression patterns using other methods (Geschwind 2000).

Microarray technology is still evolving; arrays contain many more gene probes and sensitivity is improving. The value of gene expression observations, however, will be determined by our ability to place microarray data into a biological context, and this remains a major challenge (Mirnics 2001; Becker 2001; Geschwind 2001; Miles 2001).

## Genomics and Complex Psychiatric Disorders

Major depression, bipolar disorder, autism, schizophrenia and other complex brain diseases rank among the most debilitating and costliest disorders afflictions of humankind (Hyman 2000a). The complexity of these disorders is due to a combination of the likely polygenic nature of these disorders and the essential role played by environmental perturbations in the expressed dysfunction (Weinberger 1995; Tsuang et al. 2001; Tsuang 2000; Pulver 2000; Andreasen 1996). It is understandable, therefore, that mutational analysis of single genes in different cohorts often has led to confusing and non-reproducible results. Advances in the diagnosis of disease, the application of quantitative neuroanatomical methods to postmortem tissue and the neuroimaging methods used have facilitated a greater understanding of the pathophysiology of psychiatric disorders.

The development of novel tools of functional genomics enables us to approach questions addressing the etiology of complex psychiatric disorders from a novel, hypothesis-free, data-driven angle. Thus, initial application of high-density microarrays to a particular disease eliminates the need for a preconceived hypothesis; this will not limit data collection. Transcriptome expression data can be obtained using a variety of approaches, including completely impartial hierarchical clustering to informed assessment using biological criteria. In all instances one can ask simply, "What are the specific gene expression changes associated with this disorder?" One can formulate relevant biological hypotheses *after* data acquisition, with imaginative analysis of the relationships within and across the transcriptomes that are successful in driving many discovery processes (Holter et al. 2001; Hastie et al. 2001; Tamayo et al. 1999; Golub et al. 1999; Toronen et al. 1999; Brown et al. 2000; Raychaudhuri et al. 2001; Eisen et al. 1998). As eloquently put by Eric Lander of the Whitehead Institute, microarray-based research is "… a hypothesis free search for genes, because it requires no notion in advance of how the gene might contribute to the disease. Instead, the researchers use powerful molecular biological tools to spot differences between people, or even individual cells, that have the disease and those that dont. Im a big fan of such ignorance-based techniques, because humans have a lot of ignorance, and we want to play to our strong suit." (Keystone Millenium Meeting: A Trends Guide 2000).

The first datasets from DNA arrays applied to a nervous system disorders produced a wealth of information. Using cDNA arrays with over 5,000 genes on tissue from a subject with multiple sclerosis, Whitney et al. (1999), found that in acute lesions, 62 genes were differentially expressed. These genes included the Duffy chemokine receptor, interferon regulatory factor-2, and tumor necrosis factor alpha receptor-2, among others, most of them not previously associated with the disease process of multiple sclerosis. Ginsberg and colleagues have been at the forefront of analyzing the expression profile of single cells and RNA content of tangles and plaques in Alzheimers disease (Ginsberg et al. 1997, 1999, 2000). In their recent study (Ginsberg et al. 2000), transcriptome profiling was carried out on individual normal and tangle-bearing hippocampal CA1 neurons using cDNA microarrays with >18,000 probes. The expression differences were further quantified by reverse Northern blot analysis using probes for 120 selected mRNAs on custom-made cDNA arrays. The authors uncovered a range of AD-related gene expression changes that included phosphatases/kinases, cytoskeletal proteins, synaptic proteins, glutamate receptors, and dopamine receptors. In an impressive effort by Lewohl et al. (2000, 2001), transcriptome profiling of postmortem samples of superior frontal cortex of alcoholics and non-alcoholics was performed on both cDNA and oligonucleotide arrays. These studies detected the expression of >4,000 genes in the superior frontal cortex, of which 163 were found to be differentially expressed by >40% between alcoholics and non-alcoholics, including myelin and cell cycle-related genes. Over the last several years, we have identified a set of informative transcriptome changes associated with schizophrenia (see below).

More recently, microarray studies involving postmortem human material have reported unique datasets associated with the pathophysiology of Rett syndrome (Colantuoni et al. 2001; Johnston et al. 2001), autism (Purcell et al. 2001) and brain tumors (Sallinen et al. 2000; Evans et al. 2001). These findings, combined with microarray analysis of animal models of Parkinsons Disease (Mandel et al. 2000; Grunblatt et al. 2001), ischemia (Jin et al. 2001) and drug abuse (Xie et al. 2002), strongly argue that microarrays will soon become a routinely used analytical method in all expression profiling experiments involving brain tissue. This process in the future may take the form of "gene expression voxelation", a procedure invented and first implemented by Smith and co-workers (Brown et al. 2001). In this procedure, the brain tissue to be analyzed is divided into small 3D tissue cubes called "voxels", and each of these voxels is analyzed for the expression of tens of thousands of genes using a separate microarray. The obtained data create a three-dimensional brain map for each of the genes, and the spatial distribution of expression changes creates unique patterns that can be analyzed for co-regulation using principal component analysis and other data mining tools. Furthermore, the expression analysis can be combined with DNA regulatory sequence and promoter information (Livesey et al. 2000), creating a co-regulational dataset of unparalleled power.

## Transcriptome Changes in Schizophrenia

Complex disorders such as schizophrenia will benefit most from functional genomics and novel data analysis strategies. Schizophrenia affects ~1% of the population worldwide, and ranks as one of the most costly human disorders (Hyman 2000a,b; Carpenter and Buchanan 1994). The disease typically presents with a late adolescence or early adulthood onset of symptoms, which include delusions and hallucinations (positive symptoms), decreased motivation, emotional expression and social interactions (negative symptoms), and impaired executive functions and memory (cognitive symptoms; Lewis 2000; Lewis and Lieberman 2000). Deficits in cognitive processing may be present from early childhood and appear to be directly related to the future development of the clinical syndrome (Michie et al. 2000).

Schizophrenia is one of the most complex brain disorders with multiple genetic and environmental factors implicated (Weinberger 1995; Tsuang 2000; Andreasen 1996), yet their exact contribution to disease pathogenesis remains unknown. Alterations in the brains of subjects with schizophrenia have been reported in a number of regions, including the hippocampus, superior temporal gyrus, and thalamus (Harrison 1999; McCarley et al. 1999). Most prominently, converging observations from clinical, neuroimaging, histological and anatomical studies have found disturbances in the dorsal prefrontal cortex (PFC) to be a hallmark of the disease (Volk et al. 2000; Glantz and Lewis 1997, 2000; Andreasen et al. 1997; Weinberger et al. 1996; Selemon et al. 1995; Bertolino et al. 2000). Abnormal PFC function, related to many of the cognitive disturbances in schizophrenia, may involve altered synaptic structure. PFC studies report reductions in gray matter volume (Sanfilipo et al. 2000; Goldstein et al. 1999) and increased cell packing density (Lewis and Lieberman 2000; Selemon et al. 1995, 1998)but without a change in total neuron number (Pakkenberg 1993). Decreased levels of synaptophysin (Glantz and Lewis 1997; Perrone-Bizzozero et al. 1996; Karson et al. 1996)and lower density of postsynaptic dendritic spines (Glantz and Lewis 2000; Garey et al. 1998) provide further arguments that altered presynaptic structure (and possibly function) may be a critical feature of schizophrenia. Gene expression changes related to neurotransmission (Volk et al. 2000: Akbarian et al. 1995a, 1996; Meador-Woodruff et al. 1997) and/or second messenger systems (Dean et al. 1997; Hudson et al. 1999; Shimon et al. 1998) have been observed in postmortem studies of schizophrenia. However, these studies have been limited in extent and focused on a relatively restricted number of gene products.

Whereas linkage studies have implicated more than 10 chromosomal loci in the inheritance of schizophrenia (Tsuang et al. 2001; Tsuang 2000, 2001; Pulver 2000), positional cloning efforts have failed to uncover the underlying factors of genetic vulnerability. Even if the efforts of identifying individual schizophrenia susceptibility genes become successful, it is likely that their number will be substantial. With such approaches, it may remain quite difficult to define critical molecular relationships that will help identify the best candidates for a targeted drug intervention. Expression profiling studies of postmortem tissue in schizophrenia offer several important advantages over other approaches:

1) Microarrays can perform a massive and relatively rapid screening. One can analyze the expression profile of schizophrenia using microarrays that contain >70% of the human genome. Proteins in the postmortem brain are likely to be less intact than mRNA, and large-scale analysis of them is more complex. Furthermore, microarrays analyze transcripts that are intrinsic to the harvested brain region whereas protein profiles reflect intrinsic and extrinsic sources, making it more difficult to uncover changes occurring in local cells.

2) Expression changes are potentially associated with the symptoms of schizophrenia. A gene mutation or SNPs that demonstrate heritable association with schizophrenia may not be directly linked to the symptoms of the disease. In contrast, some common expression changes will be a combined product of multiple genetic vulnerabilities and environmental factors, and these summative effects may manifest as symptoms of the disease. This strategy has been informative in understanding specific types of cancer. Uncovering conserved transcriptome changes may lead to putative drug targets with much greater

efficiency than DNA sequence abnormalities that alone may not be sufficient to produce schizophrenia.

3) We can learn about the action of antipsychotic drugs.

A limited number of expression changes will be a result of the pharmacotherapies used to treat schizophrenia. Although some gene expression differences associated with neuroleptic exposure have been reported in both rodent and primate models (Pierri et al. 1999), the effects of antipsychotic medication on global gene expression is currently not well understood. Comparisons of expression profiling of medicated and unmedicated subjects with schizophrenia, together with studies of chronic antipsychotic exposure in non-human primates, will help us understand how antipsychotic medications affect molecular cascades in specific brain regions and which changes in postmortem studies may be related to the treatment of the disease. Such information may become essential for the targeted design of more selective and potent compounds with antipsychotic effect (Kawanishi et al. 2000; Jain 2000; Zanders 2000).

## Robustly and Consistently Changed Genes

In the context of our own gene expression studies, converging observations from clinical, neuroimaging and anatomical studies focused our attention on the dorsal PFC as a major site of alterations in schizophrenia (Volk et al. 2000; Glantz and Lewis 1997, 2000; Andreasen et al. 1997; Weinberger et al. 1996; Selemon et al. 1995; Bertolino et al,. 2000). In an attempt to take advantage of the strategy using high-density microarray transcriptome profiling, we recently compared the prefrontal cortex of subjects with schizophrenia to matched controls (Middleton et al. 2002; Mirnics et al. 2000). Once we initiated data analysis, it rapidly became clear that we would have to use comprehensive analytic strategies to understand the dataset.

First, we undertook the most straightforward analytical approach by asking which genes are the most changed between the subjects with schizophrenia and their matched controls (Mirnics et al. 2000, 2001c). In this process, specific signal is separated out from noise based on more or less conservative absolute and relative signal intensity measurements. This procedure was followed by a sorting process in which we simultaneously defined the most frequently and most robustly changed genes. These two different measurements are essential, as each includes different information about the disease process. To identify the most robustly changed genes, we calculated the Z-score for each gene in each comparison between a schizophrenic and control sample (Middleton et al. 2002). This step was followed by averaging the Z-scores and ranking all the genes. Next, using inferential statistics, we established the probability of any of the observations occurring by chance. In addition, we defined the most consistently changed genes. We made the assumption that the most robust change (largest Z-score) may not be the most physiologically important change. In this analysis, the magnitude of the change was not weighted but rather the frequency of change across matched pairs of schizophrenic and control subjects.

Using these analyses, we identified several genes that showed both robust and consistent expression decreases. Not surprisingly, a significant number of genes were related to synaptic physiology: N-ethylmaleimide sensitive factor (NSF) and its attachment proteins are critical to provide energy for synaptic vesicle fusion; Synapsin II (SYN2) regulates vesicle availability for exocytosis and vesicular AT-Pase is responsible for synaptic vesicle homeostasis (Mirnics et al. 2000).

At the postsynaptic membrane, the expression of regulator of G-protein signaling 4 (RGS4) exhibited a prominent and robust decrease (Mirnics et al. 2001c). RGS4 plays a crucial role in regulating the duration of signaling many ligand-bound G-protein coupled receptors (GPCR) by accelerating the hydrolysis of GTP from a G alpha subunit, resulting in a limited duration of the G-protein coupled receptor signaling (for a review, see De Vries et al. 2000). The magnitude of the RGS4 transcript reduction in schizophrenia was similar across three neo-cortical regions (prefrontal, motor and visual cortices), with a uniform transcript reduction across all layers harboring RGS4 + cells (II-III and V-VI). Changes in RGS4 expression were not observed in subjects with major depression or in non-human primates on chronic neuroleptic medication. This finding argues that the observed decrease in RGS4 may be directly associated with schizophrenia rather than its treatment or with neuropsychiatric disorders in general. Interestingly, RGS4 maps to cytogenetic region 1q21-22, a chromosomal area recently implicated in the transmission of schizophrenia susceptibility (Brzustowicz et al. 2000), suggesting that RGS4 expression deficits may be a result of a specific DNA sequence. Indeed, our recent analysis of polymorphisms in the RGS4 gene demonstrated an increased probability of 5' SNP inheritance in patients with schizophrenia across two independent cohorts, with a third cohort exhibiting a trend in heritibility (Chowdari et al., submitted for publication). If this observation represents a general hallmark of schizophrenia, the RGS4 studies would be the first to exemplify how expression information from microarray studies can be used successfully to uncover schizophrenia susceptibility genes.

In addition, several metabolic genes showed considerable and consistent expression decreases across the prefrontal cortices of subjects with schizophrenia (Middleton et al. 2002). These included cytosolic *Malate Dehydrogenase type 1* (MAD1), *mitochondrial Glutamate-Oxaloacetate Transaminase type 2, Ornithine Decarboxylase Antizyme Inhibitor* and *Ornithine Aminotransferase*. Interestingly, when we assessed individual genes in these pathways for changes in haloperidol-treated monkeys, we observed an unexpected increase in the expression of the MAD1, raising the intriguing possibility that the antipsychotic treatment may counteract the MAD1 expression decrease observed in schizophrenia. The MAD1 expression increase was most prominent in the deep cortical layers, including layer V, which contains the highest density of dopamine D2 receptors.

From the glutaminergic transmission gene group, *glutamate receptor 2* (AMPA2) showed the largest expression decrease, whereas *glutamic acid decarboxylase 1 67kD* (GAD67) was the most robustly and consistently affected gene related to GABA-ergic neurotransmission (Mirnics et al. 2000). These findings were reported previously using more conventional methods (Volk et al. 2000; Eastwood et al. 1995).

Several expressed sequence tags (ESTs) and hypothetical proteins also reported noteworthy expression changes. At least one of these ESTs represents a new gene class with currently unknown function and a remarkable expression gradient across the neocortex (Mirnics et al. 2001d). This molecule has no significant homology to any known mammalian or invertebrate genes, yet it shows a high degree of similarity to several other ESTs in the NCBI database. We are in a process of establishing the biological function of this novel gene family.

In addition to the changes described above, we identified an additional 30-40 genes with changed expression that may be associated with schizophrenia or its treatment. Due to their relatively modest magnitude of change and low abundance, they are difficult to separate from the experimental noise. These expression reports will have to be carefully examined by more conventional methods of molecular biology.

## Changed Functional Pathways

Individual, consistently changed genes can provide clues regarding potential group changes for interrelated molecular cascades. The challenges of gene group analysis lie, in part, in the formulation of such related molecular components in a supervised fashion. This approach can reveal relationships between functional cellular networks that are not obvious from examining lists of genes or classical hierarchical analysis. In our own studies, we used a supervised approach, assuming that a single gene expression deficit in a functional pathway would change expression of the other genes belonging to the same pathway (Middleton et al. 2002; Mirnics et al. 2000).

Thus far, we have identified only a few cascades that showed a consistent expression difference between subjects with schizophrenia and matched controls. These gene groups were related to presynaptic secretory function (PSYN gene group), glutamate and GABA neurotransmission, malate shuttle, transcarboxylic acid (TCA) cycle, ornithine/polyamine, aspartate/alanine, and ubiquitin metabolism groups (Middleton et al. 2002; Mirnics et al. 2000).

The PSYN genes encode proteins that regulate the "mechanics" of neurotransmitter release, and the group was formulated subjectively from literature reports. Within this group, the expression levels of two genes, NSF and SYN2 were consistently decreased. Yet, we also observed other PSYN genes with expression decreases that were present within a minority of subjects with schizophrenia. These genes never showed an increased expression in relation to matched controls, arguing that individual variability between subjects cannot account for our findings (Mirnics et al. 2000).

Expression of gene groups involved in glutamate and the GABA neurotransmission system was also decreased, supporting previous studies that have documented changes of individual genes within these latter groups in the PFC of schizophrenic subjects (Volk et al. 2000; Akbarian et al. 1995 b,c,1996; Chen et al. 1998; Collinge and Curtis 1991; Harrison et al. 1991). The etiology of these changes remains unknown, yet it is possible that these group-wise changes are a direct adaptive response to altered presynaptic function.

The well-accepted listing of metabolic groups from the Kyoto Encyclopedia of Genes and Genomes (KEGG) formed the basis for our analysis of 70 different gene clusters (Kanehisa and Goto 2000). Remarkably, only five metabolic gene groups displayed significant alterations (Middleton et al. 2002). These included the malate shuttle, transcarboxylic acid (TCA) cycle, ornithine/polyamine, aspartate/alanine, and ubiquitin metabolism groups. When analyzed across all subjects with schizophrenia, the mean expression levels of each of the five most affected gene groups were consistently and significantly decreased compared to matched controls.

## Indivuality of Expression Changes

Individuals with schizophrenia show a continuum of clinical manifestations, and this phenotypic diversity of the disease symptoms is likely reflected at the level of gene expression changes (Mirnics and Lewis 2001). Hence, we have hypothesized that schizophrenia will sort into a continuum of molecular expression phenotypes and genotypes, albeit with significantly overlapping patterns of alterations. Our microarray analysis argues that the hallmark of schizophrenia is an impaired function in neuronal communication and not an exact genetic deficit (Mirnics et al. 2001b).

The gene group analysis allowed us to test our hypothesis that deficits in different individual genes within a functional pathway may lead to common deficits converging onto the same cellular function (Mirnics et al. 2001b). In our studies, we found that the specific members of the PSYN gene group showing decreased expression varied across the pairwise comparisons. This created different patterns of decreased gene expression in individual schizophrenic subjects. In addition, the magnitude of the decrease was different for individual PSYN genes, varying from subject to subject. Across the 10 samples from subjects with schizophrenia, at least four different patterns of decreased expression of PSYN genes emerged, regardless of control subject used for comparison (Mirnics et al. 2000). The observed variations suggest that distinct patterns of PSYN gene decreases seen on the microarrays are characteristic for subjects with schizophrenia, and that they may be a result of different genetic/environmental insults. In summary, it appears that the diverse clinical symptoms and vulnerability factors that are the hallmark of schizophrenia are also apparent at the transcriptome level (Mirnics et al. 2001b).

## Anatomical Localization of the Expression Changes

In microarray experiments, verifying gene expression differences usually is not sufficient; anatomical localization of specific transcripts allows one to identify specific cellular phenotypes that may be responsible for transcriptome changes (Mirnics et al. 2001a,b). In contrast to northern or PCR assays, in situ hybridization provides an opportunity to both verify the changes detected by microarrays and to reveal if the gene expression change is uniform across all cells expressing the transcript. For example, a 50% change reported by the microarray may be a consequence of a complete loss in half of the cells that express the transcript. Yet, it may also represent a 50% decrease in the transcript level in all cells. Furthermore, changes can be co-localized to individual cell types that may share a common function, location or anatomical projection. The interpretation of the data becomes different based on this cellular context: down-regulation of a presynaptic marker in cortical layer III and transcript loss of the glutamate receptor in layer V will have a different meaning than if both changes occur in layer IV neurons.

## Gene Expression Changes are Present in Multiple Cohorts

Replicate findings, across different subject cohorts, are the gold standard for clinical studies. This approach has been particularly difficult for a complex disease like schizophrenia. Recent microarray studies of the PFC in schizophrenia, however, demonstrate the power of this experimental approach. Changes in the expression of genes encoding multiple presynaptic markers, proteins of polyamine metabolism, ubiquination and glutamate transmitter system have been reported by several groups, across different cohorts and using different experimental methods (Vawter et al. 2001; Mirnics et al. 2000, 2001a; Vawter 2001). Furthermore, these studies identified changed transcript levels for soluble malate dehydrogenase 1, GluR2 (AMPA2), RGS4, ApoL1, ApoL2, 14-3-3, glial markers and NSF across multiple cohorts (Middleton et al. 2002; Vawter 2001; Hakak et al. 2001; Bahn 2001; Sklar 2001). In the context of our own studies, Feron et al. found that RGS4 is one of the most down-regulated genes in cultured dermal fibroblasts obtained from patients with schizophrenia (Feron 2001).

## The Emerging Model

In light of these findings, we proposed a model (Mirnics et al. 2001b) suggesting that different genetic insults related to proteins involved in synaptic communication, combined with pre- or perinatal environmental factors, may lead to the shared clinical manifestations of schizophrenia. The expression changes that we and others have described are part of a very complex disease process and may represent only the end-point of a dynamically evolving disease. If schizophrenia indeed sorts into a continuum of phenotypes and associated genotypes, the relationship between expression deficits will be essential to understand. We believe that the neurodevelopmental context of schizophrenia is characterized by a deficit that affects normal synaptic function. In our model, exuberantly produced synapses in childhood (Huttenlocher and Dabholkar 1997; Huttenlocher 1979; Bourgeois et al. 1994) may not function optimally, leading to overpruning during adolescence, which culminates in the symptoms of the disease. Why would such overpruning occur? Deficits in synaptic efficacy due to altered expression of a variety of combinations of genes encoding synaptic proteins may lead to poorly sustained synapses. Alternatively, the interplay of cellular metabolism and synaptic activity may reflect altered energy metabolism that leads to the elimination of non-optimally functioning synapses. In this context, changes in GABAergic and glutaminergic transmission may be secondary to a synaptic-metabolic deficit and are phenotype-specific manifestations of a more generic malfunction. While RGS4 down-regulation could represent an adaptational response, our initial genetic studies indicate that RGS4 deficits may lead to an imbalance that adds further susceptibility to poorly functioning synapses.

## The Future

Large datasets are readily obtained with new methodologies such as gene microarrays. Such methods afford the opportunity to share massive datasets, with precious and rare resources able to be combined and analyzed across multiple centers.

The future holds the promise for the routine analysis of transcriptome profiling of single cells or cell compartments (Eberwine 2001a,b; Eberwine et al. 2001a,b) by SAGE and microarrays; data from postmortem studies will be able to influence in a more intelligent fashion the genetic screening of candidate genes and genetic findings already are leading to targeted transgenic animal production and novel drug target developments (Kawanishi et al. 2000; Jain 2000; Zanders 2000). Furthermore, genome-wide single nucleotide polymorphism (SNP) studies can be linked to brain imaging studies and cross-referenced to postmortem gene expression findings. Ultimately, we need to analyze the individual with a disease from many angles  clinical symptoms, expression studies, proteomics data, genetic susceptibility (Lindblad-Toh et al. 2000; Sachidanandam et al. 2001; Hacia et al. 1999), functional brain imaging ( McCarley et al. 1999; Weinberger et al. 1999; Van Horn et al. 2001) and drug responsiveness (Kawanishi et al. 2000; Jain 2000). Bioinformatics will provide the ultimate strategy of creating a single, merged database that is completely cross-referenced (Sherlock et al. 2001; Masys 2001; Jenssen et al. 2001; Kellam 2001; Gardiner-Garden and Littlejohn 2001; Edgar et al. 2002). The necessary tools to pursue this strategy may now be in place.

## References

Akbarian S, Huntsman MM, Kim JJ, Tafazzoli A, Potkin SG, Bunney WE Jr, Jones EG. (1995a) GABA receptor subunit gene expression in human prefrontal cortex: comparison of schizophrenics and controls. Cereb Cortex 5:550–560

Akbarian S, Smith MA, Jones EG (1995b) Editing for an AMPA receptor subunit RNA in prefrontal cortex and striatum in Alzheimer's disease, Huntington's disease and schizophrenia. Brain Res 699:297–304

Akbarian S, Smith MA, Jones EG (1995c) Gene expression for glutamic acid decarboxylase is reduced without loss of neurons in prefrontal cortex of schizophrenics. Arch Gen Psychiatr 52:258–266

Akbarian S, Sucher NJ, Bradley D, Tafazzoli A, Trinh D, Hetrick WP, Potkin SG, Sandman CA, Bunney WE Jr, Jones EG (1996) Selective alterations in gene expression for NMDA receptor subunits in prefrontal cortex of schizophrenics. J Neurosci 16:19–30

Alizadeh AA, Eisen MB, Davis RE, Ma C, Lossos IS, Rosenwald A, Boldrick JC, Sabet H, Tran T, Yu X, Powell JI, Yang L, Marti GE, Moore T, Hudson J Jr, Lu L, Lewis DB, Tibshirani R, Sherlock G, Chan WC, Greiner TC, Weisenburger DD, Armitage JO, Warnke R, Staudt LM (2000) Distinct types of diffuse large B-cell lymphoma identified by gene expression profiling. Nature 403:503–511

Andreasen NC (1996) Pieces of the schizophrenia puzzle fall into place. Neuron 16: 697–700

Andreasen NC, O'Leary DS, Flaum M, Nopoulos P, Watkins GL, Boles Ponto LL, Hichwa RD (1997) Hypofrontality in schizophrenia: distributed dysfunctional circuits in neuroleptic-naive patients. Lancet 349: 1730–1734

Bahn S (2001) Gene expression analysis in schizophrenia: reproducible upregulation of several members of the Apolipoprotein l family located in a high susceptibility locus for schizophrenia on chromosome 22. In: Stanley Foundation, Washington, D.C., Abstracts

Barranger JM, Novelli EA (2001) Gene therapy for lysosomal storage disorders. Expert Opin Biol Ther 1: 857–867

Becker KG (2001) The sharing of cDNA microarray data. Nature Rev Neurosci 2: 438–440

Bertolino A, Esposito G, Callicott JH, Mattay VS, Van Horn JD, Frank JA, Berman KF, Weinberger DR (2000) Specific relationship between prefrontal neuronal N-acetylaspartate and activation of working memory cortical network in schizophrenia. Am J Psychiatr 157:26–33

Bertucci F, Bernard K, Loriod B, Chang YC, Granjeaud S, Birnbaum D, Nguyen C, Peck K, Jordan BR (1999) Sensitivity issues in DNA array-based expression measurements and performance of nylon microarrays for small samples. Human Mol Genet 8: 1715–1722

Bourgeois JP, Goldman-Rakic PS, Rakic P (1994) Synaptogenesis in the prefrontal cortex of rhesus monkeys. Cereb Cortex 4: 78–96

Brail LH, Jang A, Billia F, Iscove NN, Klamut HJ, Hill RP (1999) Gene expression in individual cells: analysis using global single cell reverse transcription polymerase chain reaction (GSC RT-PCR). Mutat Res 406: 45–54

Brown MP, Grundy WN, Lin D, Cristianini N, Sugnet CW, Furey TS, Ares M Jr, Haussler D (2000) Knowledge-based analysis of microarray gene expression data by using support vector machines. Proc Natl Acad Sci USA 97: 262–267

Brown VM, Ossadtchi A, Khan AH, Yee S, Lacan G, Melega WP, Cherry SR, Leahy RM, Smith DJ (2001) Multiplex three dimensional brain gene expression mapping in a mouse model of Parkinson's disease. In: Geschwind DH (ed) Society For Neuroscience Annual Meeting – DNA short course, pp. 111–124

Brzustowicz LM, Hodgkinson KA, Chow EW, Honer WG, Bassett AS (2000) Location of a major susceptibility locus for familial schizophrenia on chromosome 1q21-q22. Science 288:678–682

Carninci P, Shibata Y, Hayatsu N, Itoh M, Shiraki T, Hirozane T, Watahiki A, Shibata K, Konno H, Muramatsu M, Hayashizaki Y (2001) Balanced-size and long-size cloning of full-length, cap-trapped cDNAs into vectors of the novel lambda-FLC family allows enhanced gene discovery rate and functional analysis. Genomics 77:79–90

Carpenter WT, Buchanan RW (1994) Schizophrenia. New Engl J Med 330: 681–690

Chen AC, McDonald B, Moss SJ, Gurling HM (1998) Gene expression studies of mRNAs encoding the NMDA receptor subunits NMDAR1, NMDAR2A, NMDAR2B, NMDAR2C, and NMDAR2D following long-term treatment with cis-and trans-flupenthixol as a model for understanding the mode of action of schizophrenia drug treatment. Brain Res Mol Brain Res 54: 92–100

Cirelli C, Tononi G (1999) Differences in brain gene expression between sleep and waking as revealed by mRNA differential display and cDNA microarray technology. J Sleep Res 8 (Suppl):44–52

Colantuoni C, Jeon OH, Hyder K, Chenchik A, Khimani AH, Narayanan V, Hoffman EP, Kaufmann WE, Naidu S, Pevsner J (2001) Gene expression profiling in postmortem Rett Syndrome brain: differential gene expression and patient classification. Neurobiol Dis 8:847–865

Collinge J, Curtis D (1991) Decreased hippocampal expression of a glutamate receptor gene in schizophrenia. British J Psychiat 159:857–859

Dean B , Opeskin K, Pavey G, Hill C, Keks N (1997) Changes in protein kinase C and adenylate cyclase in the temporal lobe from subjects with schizophrenia. J Neural Transm 104:1371–1381

DeRisi J, Penland L, Brown PO, Bittner ML, Meltzer PS, Ray M, Chen Y, Su YA, Trent JM (1996) Use of a cDNA microarray to analyse gene expression patterns in human cancer. Nature Genet 14:457–460

De Vries L, Zheng B, Fischer T, Elenko E, Farquhar MG (2000) The regulator of G protein signaling family. Annu Rev Pharmacol Toxicol 40:235–271

Dongre AR, Opiteck G, Cosand WL, Hefta SA (2001) Proteomics in the post-genome age. Biopolymers 60:206–211

Eastwood SL, McDonald B, Burnet PW, Beckwith JP, Kerwin RW, Harrison PJ (1995) Decreased expression of mRNAs encoding non-NMDA glutamate receptors GluR1 and GluR2 in medial temporal lobe neurons in schizophrenia. Brain Res Mol Brain Res 29:211–223

Eberwine J (2001a) Molecular biology of axons. "A Turning Pointellipsis". Neuron 32: 959–960

Eberwine J (2001b) Single-cell molecular biology. Nature Neurosci 4 (Suppl):1155–1156

Eberwine J, Kacharmina JE, Andrews C, Miyashiro K, McIntosh T, Becker K, Barrett T, Hinkle D, Dent G, Marciano P (2001a) mRna expression analysis of tissue sections and single cells. J Neurosci 21:8310–8314

Eberwine J, Job C, Kacharmina JE, Miyashiro K, Therianos S (2001b) Transcription factors in dendrites: dendritic imprinting of the cellular nucleus. Results Probl Cell Differ 34:57–68

Edgar R, Domrachev M, Lash AE (2002) Gene expression omnibus: NCBI gene expression and hybridization array data repository. Nucleic Acids Res 30:207–210

Eisen MB, Spellman PT, Brown PO, Botstein D (1998) Cluster analysis and display of genome-wide expression patterns. Proc Natl Acad Sci USA 95:14863–14868

Evans CO, Young AN, Brown MR, Brat DJ, Parks JS, Neish AS, Oyesiku NM (2001) Novel patterns of gene expression in pituitary adenomas identified by complementary deoxyribonucleic acid microarrays and quantitative reverse transcription-polymerase chain reaction. J Clin Endocrinol Metab 86:3097–3107

Feron F (2001) Microarray studies – an integrative platform for case-control studies and animal models of schizophrenia. In: 7th Symposium on the neurovirology and neuroimmunology of schizophrenia and bipolar disorder. Stanley Foundation, Washington, D.C., Abstracts.

Freeman WM, Walker SJ, Vrana KE (1999) Quantitative RT-PCR: pitfalls and potential. Biotechniques 26: 112–122, 124–125

Garey L, Ong WY, Patel TS, Kanani M, Davis A, Mortimer AM, Barnes TR, Hirsch SR (1998) Reduced dendritic spine density on cerebral cortical pyramidal neurons in schizophrenia. J Neurol Neurosurg Psychiat 65:446–453

Gardiner-Garden M, Littlejohn TG (2001) A comparison of microarray databases. Brief Bioinform 2:143–158

Geschwind DH (2000) Mice, microarrays, and the genetic diversity of the brain. Proc Natl Acad Sci USA 97:10676–10678

Geschwind DH, Ou J, Easterday MC, Dougherty JD, Jackson RL, Chen Z, Antoine H, Terskikh A, Weissman IL, Nelson SF, Kornblum HI (2001a) A genetic analysis of neural progenitor differentiation. Neuron 29:325–339

Geschwind DH (2001b) Sharing gene expression data: an array of options. Nature Rev Neurosci 2:435–438

Ginsberg SD, Crino PB, Lee VM, Eberwine JH, Trojanowski JQ (1997) Sequestration of RNA in Alzheimer's disease neurofibrillary tangles and senile plaques. Ann Neurol 41: 200–209

Ginsberg SD, Crino PB, Hemby SE, Weingarten JA, Lee VM, Eberwine JH, Trojanowski JQ (1999) Predominance of neuronal mRNAs in individual Alzheimer's disease senile plaques. Ann Neurol 45:174–181

Ginsberg SD, Hemby SE, Lee VM, Eberwine JH, Trojanowski JQ (2000) Expression profile of transcripts in Alzheimer's disease tangle-bearing CA1 neurons. Ann Neurol 48: 77–87

Glantz LA, Lewis DA (1997) Reduction of synaptophysin immunoreactivity in the prefrontal cortex of subjects with schizophrenia. Regional and diagnostic specificity. Arch Gen Psychiat 54:943–952

Glantz LA, Lewis DA (2000) Decreased dendritic spine density of prefrontal cortical pyramidal neurons in schizophrenia. Arch Gen Psychiat 57:65–73

Goldstein JM, Goodman JM, Seidman LJ, Kennedy DN, Makris N, Lee H, Tourville J, Caviness VS Jr, Faraone SV, Tsuang MT (1999) Cortical abnormalities in schizophrenia identified by structural magnetic resonance imaging. Arch Gen Psychiat 56: 537–547

Golub TR, Slonim DK, Tamayo P, Huard C, Gaasenbeek M, Mesirov JP, Coller H, Loh ML, Downing JR, Caligiuri MA, Bloomfield CD, Lander ES (1999) Molecular classification of cancer: class discovery and class prediction by gene expression monitoring. Science 286:531–537

Grunblatt E, Mandel S, Maor G, Youdim MB (2001) Gene expression analysis in N-methyl-4-phenyl-1,2,3,6- tetrahydropyridine mice model of Parkinson's disease using cDNA microarray: effect of R-apomorphine. J Neurochem 78:1–12

Hacia JG, Fan JB, Ryder O, Jin L, Edgemon K, Ghandour G, Mayer RA, Sun B, Hsie L, Robbins CM, Brody LC, Wang D, Lander ES, Lipshutz R, Fodor SP, Collins (1999) Determination of ancestral alleles for human single-nucleotide polymorphisms using high-density oligonucleotide arrays. Nature Genet 22:164–167

Hakak Y, FS Walker Jr, Li C, Wong WH, Davis KL, Buxbaum JD, Haroutunian V, Fienberg AA (2001) Genome-wide expression analysis reveals dysregulation of myelination- related genes in chronic schizophrenia. Proc Natl Acad Sci USA 98:4746–4751

Harrison PJ (1999) The neuropathology of schizophrenia. A critical review of the data and their interpretation. Brain 122:593–624

Harrison PJ, McLaughlin D, Kerwin RW (1991) Decreased hippocampal expression of a glutamate receptor gene in schizophrenia. Lancet 337: 450–452

Hastie T, Tibshirani R, Eisen MB, Alizadeh A, Levy R, Staudt L, Chan WC, Botstein D, Brown P (2000) 'Gene shaving' as a method for identifying distinct sets of genes with similar expression patterns. Genome Biol 1:0003.0001–0003.0021

Hastie T, Tibshirani R, Botstein D, Brown P (2001) Supervised harvesting of expression trees. Genome Biol 2:0003.0001–0003.0012

Holter NS, Maritan A, Cieplak M, Fedoroff NV, Banavar JR (2001) Dynamic modeling of gene expression data. Proc Natl Acad Sci USA 98:1693–1698

Hudson C, Gotowiec A, Seeman M, Warsh J, Ross BM (1999) Clinical subtyping reveals significant differences in calcium-dependent phospholipase A2 activity in schizophrenia. Biol Psychiat 46:401–405

Husi H, Grant SG (2001) Proteomics of the nervous system. Trends Neurosci 24:259–266

Huttenlocher P (1979) Synaptic density in human frontal cortex – Developmental changes and effects of aging. Brain Res 163:195–205

Huttenlocher PR, Dabholkar AS (1997) Regional differences in synaptogenesis in human cerebral cortex. J Comp Neurol 387:167–178

Hyman SE (2000a) The genetics of mental illness: implications for practice. Bull World Health Organ 78:455–463

Hyman SE (2000b) The NIMH Perspective: Next steps in schizophrenia research. Biol Psychiat 47:1–7

Jain KK (2000) Applications of biochip and microarray systems in pharmacogenomics. Pharmacogenomics 1:289–307

Jenssen TK, Laegreid A, Komorowski J, Hovig E (2001) A literature network of human genes for high-throughput analysis of gene expression. Nature Genet 28:21–28

Jin K, Mao XO, Eshoo MW, Nagayama T, Minami M, Simon RP, Greenberg DA (2001) Microarray analysis of hippocampal gene expression in global cerebral ischemia. Ann Neurol 50:93–103

Johnston MV, Jeon OH, Pevsner J, Blue ME, Naidu S (2001) Neurobiology of Rett syndrome: a genetic disorder of synapse development. Brain Dev 23 (suppl):S206–S213

Kanehisa M, Goto S (2000) KEGG: Kyoto encyclopedia of genes and genomes. Nucleic Acids Res 28:27–30

Karson CN, Griffin WS, Mrak RE, Sturner WQ, Shillcutt S, Guggenheim FG (1996) Reduced levels of synaptic proteins in the prefrontal cortex in schizophrenia. Soc Neurosci Abstr 22:1677

Kawanishi Y, Tachikawa H, Suzuki T (2000) Pharmacogenomics and schizophrenia. Eur J Pharmacol 410:227–241

Kellam P (2001) Microarray gene expression database: progress towards an international repository of gene expression data. Genome Biol 2:Reports 4011

Lander ES, Linton LM, Birren B, Nusbaum C, Zody MC, Baldwin J, Devon K, Dewar K, Doyle M, FitzHugh W, Funke R, Gage D, Harris K, Heaford A, Howland J, Kann L, Lehoczky J, LeVine R, McEwan P, McKernan K, Meldrim J, Mesirov JP, Miranda C, Morris W, Naylor J, Raymond C, Rosetti M, Santos R, Sheridan A, Sougnez C, Stange-Thomann N, Stojanovic N, Subramanian A, Wyman D, Rogers J, Sulston J, Ainscough R, Beck S, Bentley D, Burton J, Clee C, Carter N, Coulson A, Deadman R, Deloukas P, Dunham A, Dunham I, Durbin R, French L, Grafham D, Gregory S, Hubbard T, Humphray S, Hunt A, Jones M, Lloyd C, McMurray A, Matthews L, Mercer S, Milne S, Mullikin JC, Mungall A, Plumb R, Ross M, Shownkeen R, Sims S, Waterston RH, Wilson RK, Hillier LW, McPherson JD, Marra MA, Mardis ER, Fulton LA, Chinwalla AT, Pepin KH, Gish WR, Chissoe SL, Wendl MC, Delehaunty KD, Miner TL, Delehaunty A, Kramer JB, Cook LL, Fulton RS, Johnson DL, Minx PJ, Clifton SW, Hawkins T, Branscomb E, Predki P, Richardson P, Wenning S, Slezak T, Doggett N, Cheng JF, Olsen A, Lucas S, Elkin C, Uberbacher E, Frazier M, Gibbs RA, Muzny DM, Scherer SE, Bouck JB, Sodergren EJ, Worley KC, Rives CM, Gorrell JH, Metzker ML, Naylor SL, Kucherlapati RS, Nelson DL, Weinstock GM, Sakaki Y, Fujiyama A, Hattori M, Yada T, Toyoda A, Itoh T, Kawagoe C, Watanabe H, Totoki Y, Taylor T, Weissenbach J, Heilig R, Saurin W, Artiguenave F, Brottier P, Bruls T, Pelletier E, Robert C, Wincker P, Smith DR, Doucette-Stamm L, Rubenfield M, Weinstock K, Lee HM, Dubois J, Rosenthal A, Platzer M, Nyakatura G, Taudien S, Rump A, Yang H, Yu J, Wang J, Huang G, Gu J, Hood L, Rowen L, Madan A, Qin S, Davis RW, Federspiel NA, Abola AP, Proctor MJ, Myers RM, Schmutz J, Dickson M, Grimwood J, Cox DR, Olson MV, Kaul R, Raymond C, Shimizu N, Kawasaki K, Minoshima S, Evans GA, Athanasiou M, Schultz R, Roe BA, Chen F, Pan H, Ramser J, Lehrach H, Reinhardt R, McCombie WR, de la Bastide M, Dedhia N, Blocker H, Hornischer K, Nordsiek G, Agarwala R, Aravind L, Bailey JA, Bateman A, Batzoglou S, Birney E, Bork P, Brown DG, Burge CB, Cerutti L, Chen HC, Church D, Clamp M, Copley RR, Doerks T, Eddy SR, Eichler EE, Furey TS, Galagan J, Gilbert JG, Harmon C, Hayashizaki Y, Haussler D, Hermjakob H, Hokamp K, Jang W, Johnson LS, Jones TA, Kasif S, Kaspryzk A, Kennedy S, Kent WJ, Kitts P, Koonin EV, Korf I, Kulp D, Lancet D, Lowe TM, McLysaght A, Mikkelsen T, Moran JV, Mulder N, Pollara VJ, Ponting CP, Schuler G, Schultz J, Slater G, Smit AF, Stupka E, Szustakowski J, Thierry-Mieg D, Thierry-Mieg J, Wagner L, Wallis J, Wheeler R, Williams A, Wolf YI, Wolfe KH, Yang SP, Yeh RF, Collins F, Guyer MS, Peterson J, Felsenfeld A, Wetterstrand KA, Patrinos A, Morgan MJ, Szustakowki J, de Jong P, Catanese JJ, Osoegawa K, Shizuya H, Choi S, Chen YJ (2001) Initial sequencing and analysis of the human genome. Nature 409:860–921

Lewohl JM, Dodd PR, Mayfield RD, Harris RA (2000) Gene expression in human alcoholism: microarray analysis of frontal cortex. Alcohol Clin Exp Res 24:1873–1882

Lewohl JM, Wang L, Miles MF, Zhang L, Dodd PR, Harris RA (2001) Application of DNA microarrays to study human alcoholism. J Biomed Sci 8:28–36

Lewis DA (2000) Is there a neuropathology of schizophrenia? The Neurosci 6:208–218

Lewis D, Lieberman J (2000) Catching up with schizophrenia: natural history and neurobiology. Neuron 28:325–334

Lindblad-Toh K, Winchester E, Daly MJ, Wang DG, Hirschhorn JN, Laviolette JP, Ardlie K, Reich DE, Robinson E, Sklar P, Shah N, Thomas D, Fan JB, Gingeras T, Warrington J, Patil N, Hudson TJ, Lander ES (2000) Large-scale discovery and genotyping of single-nucleotide polymorphisms in the mouse. Nature Genet 24:381–386

Lipshutz RJ, Fodor SP, Gingeras TR, Lockhart DJ (1999) High density synthetic oligonucleotide arrays. Nature Genet 21 (Suppl):20–24

Livesey FJ, Furukawa T, Steffen MA, Church GM, Cepko CL (2000) Microarray analysis of the transcriptional network controlled by the photoreceptor homeobox gene Crx. Curr Biol 10:301–310

Lockhart DJ, Barlow C (2001a) Neural gene expression analysis using DNA arrays. In: Chin H, Moldin S (eds) Methods in genomic neuroscience CRC Press LLC, Boca Raton, Fl, pp 143–171

Lockhart DJ, Barlow C (2001b) Expressing what's on your mind: DNA arrays and the brain. Nature Rev Neurosci 2:63–68

Lockhart DJ, Dong H, Byrne MC, Follettie MT, Gallo MV, Chee MS, Mittmann M, Wang C, Kobayashi M, Horton H, Brown EL (1996) Expression monitoring by hybridization to high-density oligonucleotide arrays. Nature Biotechnol 14:1675–1680

Luo Z, Geschwind DH (2001) Microarray applications in neuroscience. Neurobiol Dis 8:183–193

Mandel S, Grunblatt E, Youdim M (2000) cDNA microarray to study gene expression of dopamin-ergic neurodegeneration and neuroprotection in MPTP and 6-hydroxydopamine models: implications for idiopathic Parkinson's disease. J Neural Transm 60 (Suppl):117–124

Marcotte ER, Srivastava LK, Quirion R (2001) DNA microarrays in neuropsychopharmacology. Trends Pharmacol Sci 22:426–436

Masys DR (2001) Linking microarray data to the literature. Nature Genet 28:9–10

McCarley R, Wible CG, Frumin M, Hirayasu Y, Levitt JJ, Fischer IA, Shenton ME (1999) MRI anatomy of schizophrenia. Biol Psychiat 45:1099–1119

Meador-Woodruff J, Haroutunian V, Powchik P, Davidson M, Davis KL, Watson SJ. (1997) Dopamine receptor transcript expression in striatum and prefrontal occipital cortex. Arch Gen Psychiat 54:1089–1095

Michie PT, Kent A, Stiensstra R, Castine R, Johnston J, Dedman K, Wichmann H, Box J, Rock D, Rutherford E, Jablensky A (2000) Phenotypic markers as risk factors Aust N Z J Psychiatry 34 (Suppl):S74–S85

Middleton FA, Mirnics K, Pierri JN, Lewis DA, Levitt P (2002) Gene expression profiling reveals alterations of specific metabolic pathways in schizophrenia. J Neurosci 22:2718–29@References = Miles MF (2001) Microarrays: lost in a storm of data? Nature Rev Neurosci 2:441–443

Mirnics K (2001) Microarrays in brain research: the good, the bad and the ugly. Nature Rev Neurosci 2:444–447

Mirnics K (2002) Microarrays in brain research: data quality and limitations. Curr Genomics 3, in press

Mirnics K, Lewis DA (2001) Genes and subtypes of schizophrenia. Trends Mol Med 7: 281–283

Mirnics K, Middleton FA, Marquez A, Lewis DA, Levitt P (2000) Molecular characterization of schizophrenia viewed by microarray analysis of gene expression in prefrontal cortex. Neuron 28:53–67

Mirnics K, David A, Lewis DA, Levitt P (2001a) DNA microarrays and Human brain disorders. In: Chin H, Moldin, SO (eds) Methods in genomic neuroscience. CRC Press LLC, Boca Raton, Fl, pp 171–204

Mirnics K, Middleton FA, Lewis DA, Levitt P (2001b) Analysis of complex brain disorders with microarrays: schizophrenia as a disease of the synapse. Trends Neurosci 24: 479–486

Mirnics K Middleton FA, Stanwood GD, Lewis DA, Levitt P (2001c) Disease-specific changes in regulator of G-protein signaling 4 (RGS4) expression in schizophrenia. Mol Psychiat 6: 293–301

Mirnics K, Middleton FA, Peng L, Lewis DA, Levitt P (2001d) Altered expression of a novel gene in the prefrontal cortex of subjects with schizophrenia. Soc Neurosci, Abstract n° 571.10

Olivier M, Aggarwal A, Allen J, Almendras AA, Bajorek ES, Beasley EM, Brady SD, Bushard JM, Bustos VI, Chu A, Chung TR, De Witte A, Denys ME, Dominguez R, Fang NY, Foster BD, Freudenberg RW, Hadley D, Hamilton LR, Jeffrey TJ, Kelly L, Lazzeroni L, Levy MR, Lewis SC, Liu X, Lopez FJ, Louie B, Marquis JP, Martinez RA, Matsuura MK, Misherghi NS, Norton JA, Olshen A, Perkins SM, Perou AJ, Piercy C, Piercy M, Qin F, Reif T, Sheppard K, Shokoohi V, Smick GA, Sun WL, Stewart EA, Fernando J, Tejeda, Tran NM, Trejo T, Vo NT, Yan SC, Zierten

DL, Zhao S, Sachidanandam R, Trask BJ, Myers RM, Cox DR (2001) A high-resolution radiation hybrid map of the human genome draft sequence. Science 291: 1298–1302

Pakkenberg B (1993) Total nerve cell number in neocortex in chronic schizophrenics and controls estimated using optical dissectors. Biol Psychiat 11:768–772

Perrone-Bizzozero N, Sower AC, Bird ED, Benowitz LI, Ivins KJ, Neve RL (1996) Levels of the growth-associated protein GAP-43 are selectively increased in association cortices in schizophrenia. Proc Natl Acad Sci USA 93:14182–14187

Pfeifer A, Verma IM (2001) Gene therapy: promises and problems. Annu Rev Genomics Human Genet 2:177–211

Pierri JN, Chaudry AS, Woo TU, Lewis DA (1999) Alterations in chandelier neuron axon terminals in the prefrontal cortex of schizophrenic subjects. Am J Psychiat 156:1709–1719

Pulver AE (2000) Search for schizophrenia susceptibility genes. Biol Psychiat 47:221–230

Purcell AE, Jeon OH, Zimmerman AW, Blue ME, Pevsner J (2001) Postmortem brain abnormalities of the glutamate neurotransmitter system in autism. Neurology 57:1618–1628

Raychaudhuri S, Sutphin PD, Chang JT, Altman RB (2001) Basic microarray analysis: grouping and feature reduction. Trends Biotechnol 19:189–193

Sachidanandam R, Weissman D, Schmidt SC, Kakol JM, Stein LD, Marth G, Sherry S, Mullikin JC, Mortimore BJ, Willey DL, Hunt SE, Cole CG, Coggill PC, Rice CM, Ning Z, Rogers J, Bentley DR, Kwok PY, Mardis ER, Yeh RT, Schultz B, Cook L, Davenport R, Dante M, Fulton L, Hillier L, Waterston RH, McPherson JD, Gilman B, Schaffner S, Van Etten WJ, Reich D, Higgins J, Daly MJ, Blumenstiel B, Baldwin J, Stange-Thomann N, Zody MC, Linton L, Lander ES, Altshuler D (2001) A map of human genome sequence variation containing 1.42 million single nucleotide polymorphisms. Nature 409: 928–933

Sallinen PK, Haapasalo HK, Helin HJ, Helen PT, Schraml P, Kallioniemi OP, Kononen J (2000) Identification of differentially expressed genes in human gliomas by DNA microarray and tissue chip techniques. Cancer Res 60:6617–6622

Sandberg R, Yasuda R, Pankratz DG, Carter TA, Del Rio JA, Wodicka L, Mayford M, Lockhart DJ, Barlow C (2000) From the cover: regional and strain-specific gene expression mapping in the adult mouse brain. Proc Natl Acad Sci USA 97:11038–11043

Sanfilipo M, Lafargue T, Rusinek H, Arena L, Loneragan C, Lautin A, Feiner D, Rotrosen J, Wolkin A (2000) Volumetric measure of the frontal and temporal lobe regions in schizophrenia: Relationship to negative symptoms. Arch Gen Psychiat 57: 471–480

Schena M, Shalon D, Davis RW, Brown PO (1995) Quantitative monitoring of gene expression patterns with a complementary DNA microarray. Science 270:467–470

Selemon LD, Rajkowska G, Goldman-Rakic PS (1995) Abnormally high neuronal density in the schizophrenic cortex. A morphometric analysis of prefrontal area 9 and occipital area 17. Arch Gen Psychiat 52: 805–818

Selemon LD, Rajkowska G, Goldman-Rakic PS (1998) Elevated neuronal density in prefrontal area 46 in brains from schizophrenic patients: application of a three-dimensional, stereologic counting method. J Comp Neurol 392:402–412

Sherlock G, Hernandez-Boussard T, Kasarskis A, Binkley G, Matese JC, Dwight SS, Kaloper M, Weng S, Jin H, Ball CA, Eisen MB, Spellman PT, Brown PO, Botstein D, Cherry JM (2001) The Stanford Microarray Database. Nucleic Acids Res 29:152–155

Shimon H, Sobolev Y, Davidson M, Haroutunian V, Belmaker RH, Agam G (1998) Inositol levels are decreased in postmortem brain of schizophrenic patients. Biol Psychiat 44: 428–432

Sklar P (2001) Microarray analysis of the Stanley brain collection. In: 7th Symposium on the neurovirology and neuroimmunology of schizophrenia and bipolar disorders, Stanley Foundation, Washington, D.C., Abstracts.

Southern E, Mir K, Shchepinov M (1999) Molecular interactions on microarrays. Nature Genet 21 (Suppl 1):5–9

Stoll D, Templin MF, Schrenk M, Traub PC, Vohringer CF, Joos TO (2002) Protein microarray technology. Front Biosci 7:C13–C32

Tamayo P, Slonim D, Mesirov J, Zhu Q, Kitareewan S, Dmitrovsky E, Lander ES, Golub TR (1999) Interpreting patterns of gene expression with self-organizing maps: methods and application to hematopoietic differentiation. Proc Natl Acad Sci USA 96: 2907–2912

Toronen P, Kolehmainen M, Wong G, Castren E (1999) Analysis of gene expression data using self-organizing maps. FEBS Lett 451:142–146

Tsuang M (2000) Schizophrenia: Genes and environment. Biol Psychiat 47:210–220

Tsuang MT (2001) Defining alternative phenotypes for genetic studies: what can we learn from studies of schizophrenia? Am J Med Genet 105:8–10

Tsuang MT, Stone WS, Faraone SV (2001) Genes, environment and schizophrenia. Br J Psychiat 40 (Suppl):S18–S24

Van Horn JD, Grethe JS, Kostelec P, Woodward JB, Aslam JA, Rus D, Rockmore D, Gazzaniga MS (2001) The functional magnetic resonance imaging data center (fMRIDC): the challenges and rewards of large-scale databasing of neuroimaging studies. Philos Trans R Soc Lond B Biol Sci 356: 1323–1339

Vawter MP (2001) Preliminary screening of gene expression in schizophrenia by cDNA microarray. In: 7th Symposium on the neurovirology and neuroimmunology of schizophrenia and bipolar disorder., Stanley Foundation, Washington, D.C., Abstracts

Vawter MP, Grethe JS, Kostelec P, Woodward JB, Aslam JA, Rus D, Rockmore D, Gazzaniga MS (2001) Application of cDNA microarrays to examine gene expression differences in schizophrenia. Brain Res Bull 55:641–650

Velculescu VE, Zhang L, Vogelstein B, Kinzler KW (1995) Serial analysis of gene expression. Science 270:484–487

Volk DW, Austin MC, Pierri JN, Sampson AR, Lewis DA (2000) Decreased GAD67 mRNA expression in a subset of prefrontal cortical GABA neurons in schizophrenia. Arch Gen Psychiat 57:237–45

Weinberger DR (1995) From neuropathology to neurodevelopment. Lancet 346:552–557

Weinberger DR, Mattay V, Callicott J, Kotrla K, Santha A, van Gelderen P, Duyn J, Moonen C, Frank J (1996) fMRI applications in schizophrenia research. Neuroimage 4: S118–S126

Whitney LW, Becker KG, Tresser NJ, Caballero-Ramos CI, Munson PJ, Prabhu VV, Trent JM, McFarland HF, Biddison WE (1999) Analysis of gene expression in mutiple sclerosis lesions using cDNA microarrays. Ann Neurol 46:425–428

Xie T, Tong L, Barrett T, Yuan J, Hatzidimitriou G, McCann UD, Becker KG, Donovan DM, Ricaurte GA (2002) Changes in gene expression linked to methamphetamine-induced dopaminergic neurotoxicity. J Neurosci 22:274–283

Zanders ED (2000) Gene expression analysis as an aid to the identification of drug targets. Pharmacogenomics 1:375–384

# Genetically Altered Mice as Models for Understanding Brain Disorders

*Marc G. Caron*[1] *and Raul R. Gainetdinov*[1]

## Summary

In an attempt to understand the potential contribution of various neurotransmitter systems to the elaboration of symptoms associated with central nervous systems disorders in humans, we have studied genetically modified animal models. A mouse in which the dopamine transporter gene (DAT-KO) has been inactivated provides a model of hyperdopaminergia. This mouse displays phenotypes of hyperactivity, impaired cognitive tasks, sensorimotor gating, habituation and attention. These behavioral impairments can be corrected by antipsychotic drugs and as such recapitulate some manifestations of psychotic behaviors. However, the potent calming effect of psychostimulants on the hyperlocomotion phenotype of these mice suggests some relevance to symptoms associated with attention deficit hyperactivity disorder (ADHD). In the DAT-KO mice, our results reveal an interplay between the dopamine, glutamate and serotonin systems in controlling these behaviors. A mouse that carries a hypomorphic allele of the NR1 subunit of the NMDA receptor provides a model for a hypofunctioning glutamate system. NR1 mutant mice display behavioral abnormalities that are consistent with other pharmacological models of psychotic disorders (PCP and MK-801 intoxication), and these behaviors can be ameliorated more effectively by atypical rather than by typical antipsychotics. Thus, animals in which specific genes have been genetically modified can provide valuable models that can mimic certain disease symptoms.

## Introduction

In the brain the major neurotransmitter systems involved in the function of the cortico-basal ganglia circuits include the dopaminergic, glutamatergic and GABAergic systems (Carlsson et al. 2001, Greengard 2001). These systems are implicated in the control of locomotion, cognition and affect. Dysregulation in the dopaminergic and glutamatergic systems, in particular, have been postulated

---

[1] Howard Hughes Medical Institute and Department of Cell Biology, Duke University Medical Center, Durham, NC 27710, USA

Mallet/Christen
Neurosciencess at the Postgenomic Era
© Springer-Verlag Berlin Heidelberg 2003

to contribute to several central nervous system (CNS) disorders (Carlsson et al. 2001, Greengard 2001). To gain an understanding of the role of these neuronal systems in the elaboration of physiological or pathophysiological conditions, the use of a genetic approach in the mouse has become a valuable asset. By creating animals in which the genes for certain key components of these neurotransmitter systems have been deleted or suppressed by homologous recombination, it is possible to recapitulate certain behavioral manifestations associated with various disorders (Smithies 1993). Most disorders of the CNS likely result from complex traits to which mutations in multiple genes as well as various environmental factors must contribute. Developing animal models in which the function of a single gene is genetically manipulated is unlikely to reproduce the full spectrum of a complex disorder such as schizophrenia or attention deficit hyperactivity disorder (Gainetdinov et al. 2001b). Nonetheless, these models offer experimental paradigms that are not usable for studies in humans.

In this essay, we summarize approaches our laboratory has taken using mouse genetics to elucidate how various neurotransmitter systems in the CNS might contribute to behavioral manifestations and symptoms associated with schizophrenia and attention deficit hyperactivity disorder (ADHD). We conclude with a short summary of other genetically altered mouse models with potential relevance to these conditions.

There is growing evidence that, neurobiologically, multiple neurotransmitter systems may be involved in the manifestation of schizophrenia-related behaviors. Imbalances not only in dopaminergic but also in serotonergic and glutamatergic systems within the frontostriatal circuitry are currently considered to be primarily responsible for behaviors associated with schizophrenia (Carlsson et al. 2001; Lewis and Lieberman 2000). Based on pharmacological observations, two major hypotheses have been proposed to explain psychotic behaviors. The dopamine (DA) hypothesis of schizophrenia suggests that either an excess of DA or an increased sensitivity to the neurotransmitter is the underlying pathological mechanism (Creese et al. 1976; Seeman 1987; Laruelle et al. 1999). A competing notion, the hypoglutamatergic hypothesis of schizophrenia, suggests that decreased glutamatergic transmission underlies schizophrenia, based on the ability of NMDA receptor antagonists to recapitulate some of the symptoms of schizophrenia (Tamminga 1998; Coyle 1996; Carlsson et al. 2001). These hypotheses can be examined in animals created by gene targeting that display alterations in the critical systems and brain regions associated with schizophrenia. The DA transporter knockout (DAT-KO) mice (Giros et al. 1996) represent a non-pharmacologically manipulated model of persistent hyperdopaminergia (Jones et al. 1998a; Gainetdinov et al. 1999a) and therefore are appropriate substrates to evaluate the DA hypothesis. The recently characterized NMDA receptor-deficient (NR1-KD) mice (Mohn et al. 1999) can be used to evaluate the hypoglutamatergic hypothesis because these animals have reduced NMDA receptor-mediated neurotransmission.

ADHD is a common disorder that appears mostly in school-aged children and manifests as impulsivity, hyperactivity and inattention (Barkley 1990). The cause and pathophysiology of ADHD are unknown, but compelling evidence suggests an involvement of genetic factors. Dopamine is believed to play a major role in

ADHD, but a role for norepinephrine and serotonin systems has also been sus-pected. The most commonly used pharmacotherapy for ADHD is based on the paradoxical ability of psychostimulants to produce an ameliorating (calming) effect; however, antidepressants can also show therapeutic benefits (Barkley 1990). Studies have reported an association between a polymorphism in the non-coding regions of DAT gene and ADHD (Cook et al. 1995; Gill et al. 1997; Wald-man et al. 1998). Although the functional consequences of this association are unknown, it suggests that alterations in DAT-mediated processes might contrib-ute to the pathogenesis of this disorder. Several lines of evidence suggest that the observations gained in DAT-KO mice may also be relevant for this disorder (Gainetdinov et al. 1999b, 2001b; Gainetdinov and Caron 2000).

## Genetic Animal Models for Schizophrenia-related Behaviors

For decades investigators have used pharmacological treatment of animals in an attempt to mimic symptoms associated with CNS disorders. For schizophrenia, agents that increase DA-mediated behaviors or promote behaviors associated with decreased glutamatergic neurotransmission have been used. Amphetamine is a potent agent that acts through plasma membrane and vesicular monoamine transporters to release monoamines, in particular DA, in the extracellular space (Seiden et al. 1993; Sulzer et al. 1995; Jones et al. 1998b). As such amphetamine profoundly affects motor activity, sensorimotor function, attention, aggressive behaviors, learning and memory as well as some other behaviors (Seiden et al. 1993). Chronic exposure to amphetamine mimics some symptoms of schizophre-nia in normal individuals and exacerbates symptoms in patients (Ellinwood 1967; Snyder 1973; Kokkinidis and Anisman 1981; Swerdlow and Geyer 1998). In experimental animals several behavioral correlates are primarily considered as reflections of certain psychotic manifestations. For example, amphetamine in-duces extreme hyperactivity and stereotypy in rodents that are believed to model the positive symptoms of schizophrenia. Reversal of these behaviors is com-monly used to predict potential antipsychotic activity of a drug in pre-clinical studies (Kokkinidis and Anisman 1981; Martin et al. 1998). Another behavior that is reversible by neuroleptics is the amphetamine-induced disruption of nor-mal sensorimotor gating (Swerdlow and Geyer 1998). Amphetamine treatment in animal models of schizophrenia can reliably predict DA receptor antagonism and potential efficacy of novel drugs on positive symptomatology. However, this approach is limited by its inability to model negative symptoms of the disease (Ellinwood 1967; Snyder 1973; Kokkinidis and Anisman 1981; Carlsson et al. 2001).

Glutamate neurotransmission has been implicated in schizophrenia in large part due to the psychomimetic effects of NMDA receptor antagonists (Luby et al. 1959; Javitt and Zukin 1991). In clinical investigations it has been noted that PCP and ketamine can produce psychosis in healthy humans that is similar to some symptoms of schizophrenia. In addition, the capacity of these drugs to worsen symptoms in patients with this disorder was also generally observed (Tamminga 1998; Coyle 1996; Carlsson et al. 2001). The cluster of symptoms produced by ketamine and PCP in healthy subjects includes positive symptoms (delusions and hallucinations), disorganization symptoms (including several aspects of formal thought disorder), negative symptoms (emotional blunting, social withdrawal and apathy), and cognitive impairment (deficits in attention and working memory). These observations provide the basis for the use of PCP and related drugs in animal models of schizophrenia. In rodents, PCP and other NMDA receptor antagonists are commonly used to induce schizophrenia-related behaviors that include, but are not limited to, increases in locomotion and stereotyped behavior, deficits in social interactions and disrupted sensorimotor gating and cognitive functions (Lehmann-Masten and Geyer 1991; Corbett et al. 1995). Atypical antipsychotics, which have a significant serotonergic component of action, are more effective in this model than typical antipsychotics. In particular, it was established that 5-HT2 receptor, rather than D2 DA receptor antagonism, is critical for the drug efficacy in antagonizing these behaviors (Martin et al. 1998; Meltzer 1999; Aghajanian and Marek 2000).

## Hyperdopaminergic mice – DATKO mice

In keeping with the widespread use of amphetamine as a pharmacological model of schizophrenia, a clear candidate for a genetic model of this disorder is the DA transporter knockout mouse (DAT-KO; Giros et al. 1996). Mice lacking DAT are unable to re-uptake released DA, and the neurotransmitter levels in extracellular fluid remain persistently elevated (Jones et al. 1998a; Gainetdinov et al. 1998). While amphetamine causes a transient elevation in synaptic DA, the genetic removal of DAT leads to persistent hyperdopaminergia, a result that is more consistent with the DA hypothesis of schizophrenia.

The removal of DAT leads to several physiological changes (Jones et al. 1998a; Gainetdinov et al. 1998). There is a five-fold increase in the extracellular levels of DA, as measured by in vivo quantitative microdialysis. There is a 300-fold increase in amount of time required to clear DA from the synapse, as measured by cyclic voltammetry of striatal slices. Heterozygous mice, with a single functional copy of DAT and half the normal levels of DAT protein, display an intermediate hyperdopaminergia with a two-fold elevation of extracellular DA. In response to the elevated DA levels, DA D1 and D2 (but not D3) DA receptor numbers are decreased (Giros et al. 1996; Fauchey et al. 2000). Functional coupling of DA autoreceptors was found to be lost in DAT-KO mice (Jones et al. 1999). Surprisingly, however, some populations of postsynaptic DA receptors appear to be supersensitive (Gainetdinov et al. 1999a; Fauchey et al. 2000). As a result of a lack of recycled DA and diminished autoreceptor function, DA synthesis is increased, but DA synthesis is not able to restore normal levels of intracellular DA (Jones et al. 1998a; Gainetdinov et al. 1998).

The DAT-KO mice gain weight more slowly than heterozygote and wild type mice. Females lacking the DAT show an inability to lactate and an impaired ability to care for their offspring (Giros et al. 1996). Deletion of the DAT results in anterior pituitary hypoplasia and numerous alterations in the parameters of the hypothalamo-pituitary axis (Bosse et al. 1997). These phenotypes all point to the effects of hyperdopaminergia in the pituitary. In behavioral paradigms, DAT-KO mice demonstrate a marked environment-dependent hyperactivity as revealed by enhanced locomotion, rearing and stereotypical behaviors (Giros et al. 1996; Sora et al. 1998; Gainetdinov et al. 1999b; Spielewoy et al. 2000). Importantly, the locomotion of these mice is dependent on intact dopaminergic transmission, since acute inhibition of dopamine synthesis or blockade of DA receptors by antagonists (haloperidol and clozapine) completely reverses the hyperactivity of these mice (Gainetdinov et al. 1999b; Spielewoy et al. 2000; Ralph et al. 2001). Psychostimulants such as cocaine and amphetamine produce paradoxical inhibitory effects on locomotion in DAT-KO mice (Gainetdinov et al. 1999b, 2001b; Spielewoy et al. 2001). At the same time, cocaine was still found to be rewarding in these mice (Rocha et al. 1998).

Several serotonergic drugs, but not an inhibitor of norepinephrine uptake, nisoxetine, also ameliorated hyperactivity in these mice, suggesting an important role of serotonin in the control of high dopamine-induced hyperactivity (Gainetdinov et al. 1999b). These mice also showed significant cognitive impairment in an eight-arm radial maze test, a standard approach to evaluate spatial cognitive function in rodents. In particular, in this maze test, mutant animals make significantly more perseverative errors, suggesting that these mice might display poor behavioral inhibition (Gainetdinov et al. 1999b). In addition, a marker of sensorimotor gating, pre-pulse inhibition, was found to be remarkably disrupted but could be restored with the D2 dopamine receptor antagonist raclopride (Ralph et al. 2001). Furthermore, a perseverative pattern of locomotor activity was observed in DAT-KO mice (Ralph et al. 2001). Importantly, however, no deficits in social interaction were found in these animals (Spielewoy et al. 2000). This distinctive phenotype of DAT-KO mice, resulting from persistent hyperdopaminergia, raises an important question: can these mice be considered to be a model for psychiatric diseases involving hyperdopaminergia? And if this is the case, which disorder is better recapitulated by disruption of DAT?

The dopaminergic hypothesis of schizophrenia and the well-known psychomimetic actions of amphetamine may suggest that alterations in DAT-mediated functions can contribute to this disorder. In fact DAT-KO mice reproduce several features of the amphetamine model of schizophrenia. Mutants are hyperactive and stereotypic and show significant deficits in sensorimotor gating and spatial cognitive function (Giros et al. 1996; Gainetdinov et al. 1999b; Spielewoy et al. 2000; Ralph et al. 2001). The specific type of spatial cognitive dysfunction was manifested as perseverative errors in several tasks (Gainetdinov et al. 1999b; Ralph et al. 2001). This perseverative type of cognitive dysfunction has been described in amphetamine-treated animals (Kokkinidis and Anisman 1981). Many of these behavioral manifestations of amphetamine intoxication in animals were suggested to be reflections of the positive symptoms associated with schizophrenia (Ellinwood 1967; Snyder 1973). Accordingly, the behaviors of DAT mutant mice may have the same level of validity as a model of schizophrenia as amphetamine-induced behaviors, with the advantage that genetically modified mice do not have the potential complication of drug treatment. However, similar criticisms may apply to both models. For example, several features of the DAT-KO mice do not correlate with those of individuals with schizophrenia, most notably the lack of deficit in social interactions. In addition there are remarkable alterations in DA neuron homeostasis in DAT-KO mice (Jones et al. 1998a), whereas similar major defects in DA neurochemistry have not been detected in schizophrenia patients. In linkage studies no association between markers in the DAT gene and this disorder has been found (Bodeau-Pean 1995; Inada et al. 1996; Persico and Macciardi 1997). Thus, it seems unlikely that major DAT alterations could be involved in this disorder. However, it is still possible that diminished DAT function can amplify disturbances in other neuronal components, resulting in presumably positive symptoms of schizophrenia (Persico and Macciardi 1997). Therefore, it is reasonable to conclude that DAT-KO mice may be a valuable model for studying some of the specific characteristic features of schizophrenia but may not completely reproduce the full spectrum of schizophrenia-related behaviors (Gainetdinov et al. 2001b). The reasons to believe that the DAT-KO mice might also be relevant for another disorder associated with enhanced dopaminergic tone  ADHD – will be described below.

## Hypoglutamatergic mice – NMDA receptor knockdown mice

Just as DAT-Ko mice represent a potentially useful genetic model for studying the dopamine hypothesis of schizophrenia, mice deficient in glutamatergic transmission may be useful foro studying the glutamate hypothesis. Mice deficient in NMDA glutamate receptors have been generated by hypomorphic mutation of the NR1 subunit gene and have been initially characterized in this context (Mohn et al. 1999).

Mutant mice bearing an intronic insertion of *neo*, a selectable marker gene, have reduced levels of functional NR1 message and protein, and NMDA receptor number is reduced to 10% of normal based on [$^3$H]MK-801 binding. These mice, designated here as NR1 knockdown (NR1-KD) mice, survive to adulthood. This is in contrast to NR1 null mice, which show perinatal lethality (Forrest et al. 1994; Li et al. 1994). NR1-KD mice have a reduced body weight compared to littermates but can achieve weights within the normal range.

Phenotypic characterization of NR1-KD mice revealed several similarities to pharmacological models of schizophrenia. Locomotor activity and stereotypy, assessed by digital activity monitors, were elevated in a manner similar to those seen with PCP and MK-801 administration. Both antipsychotic drugs haloperidol and clozapine could attenuate this hyperactivity, but only clozapine was effective at doses that did not affect wild type activity. The doses of both clozapine and haloperidol that were required to attenuate hyperactivity of NR1-KD mice were remarkably similar to those required for attenuation of PCP and MK-801-induced hyperactivity (Lehmann-Masten and Geyer 1991; Corbett et al. 1995; Martin et al. 1998). Additionally, the NR1-KD mice demonstrated deficiencies in social interaction that were improved with clozapine treatment. In this respect, too, NR1-KD mice were similar to PCP- or MK-801-treated mice, which display deficits in social interaction that are improved by clozapine (Corbett et al. 1995).

These observations suggest that NR1-KD mice should serve as a genetic counterpart to PCP or MK-801 models of schizophrenia. The observed increase in motor activity, deficits in social interaction and responsiveness of these behaviors to antipsychotic drugs support a model in which decreases in NMDA receptor activity can be a primary lesion in the development of both positive and negative symptoms of schizophrenia (Tamminga 1998; Coyle 1996; Carlsson et al. 2001). The other salient characteristics of schizophrenia, such as deficits in sensorimotor gating, which are reproducible in pharmacological models of schizophrenia in preliminary experiments, also appear to be present in these mice (A. Laakso, A.R. Mohn, R.R. Gainetdinov and M.G. Caron, unpublished results). These mice will be useful not simply as a genetic counterpart but as a tool for understanding the interaction between the glutamate and other neurotransmitter systems with respect to schizophrenia-related behaviors. For example, the observed hyperactivity of NR1-KD mice may have been predicted to be due to an elevation of DA transmission. However, neurochemical characterization of these mice demonstrated that striatal DA content and extracellular levels are normal (Mohn et al. 1999). These results support a proposed mechanism of glutamate-DA interactions (Javitt and Zukin 1991; Carlsson et al. 2001) that is depicted in Figure 1.

Although the genetic mutation is limited to NMDA receptors, an impact on the other neurotransmitter systems is suggested by the results of clozapine and haloperidol administration. While haloperidols selective blockade of D2 dopaminergic transmission can suppress hyperlocomotion and stereotypy, these behaviors are more effectively suppressed by clozapine, which has mixed actions on several neurotransmitter systems, in particular as a potent antagonist at 5-HT2A receptors (Martin et al. 1998). Beyond 5-HT2A receptors several other key molecules could be potentially targeted to ameliorate abnormal manifestations resulting from this type of frontostriatal dysregulation (Moghaddam and Adams 1998). Thus, these mice could be a valuable tool for delineating the circuitry

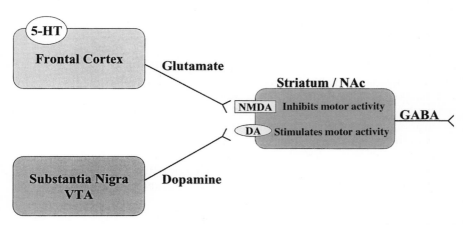

**Fig.1.** Diagram representing a simplified model of serotonin/glutamate/DA interaction in the control of locomotor activity. 5-HT, serotonin; VTA,ventral tegmental area; Nac,nucleus accumbens.

affected and serve as a test model to explore novel pharmacological means to counteract glutamatergic deficiency.

## Dopamine Transporter-deficient mice as a potential model of ADHD

As noted above, ADHD is characterized by impulsivity, hyperactivity and inattention (Barkley 1990). Interestingly, DAT-KO mice are hyperactive, especially when exposed to a novel environment (Giros et al. 1996; Gainetdinov et al. 1999b). This hyperactivity is presumably the result of a five-fold elevation in extracellular DA concentration in the brain, since the phenotype can be reversed by antagonizing D2 DA receptors (Gainetdinov et al. 1999b; Spielewoy et al. 2000). However, this novelty-driven hyperactivity occurs without a corresponding further rise in DA, suggesting that these behavioral changes might be regulated through systems other than DA (Gainetdinov et al. 1999b). Additionally, these mice are impaired in spatial cognitive tasks and behavioral assessments of information gating, habituation, attention and memory processes (Gainetdinov et al. 1999b; Rodriguiz et al. 2001). Many of these deficits can be ameliorated by various pharmacological agents used in the management of similar deficits in human conditions (Rodriguiz et al. 2001; Gainetdinov et al. 2002). As an example, DAT-KO mice show a marked decrease in locomotion in response to psychostimulants such as cocaine, amphetamine and methylphenidate, and this effect depends on enhanced serotoninergic transmission (Gainetdinov et al. 1999b). Indeed, following their exposure to a novel environment, treatment of DAT-KO mice with compounds that raise or mimic serotonin, such as fluoxetine, quipazine or precursors of serotonin, completely recapitulated the effects of psychostimulants on hyperactivity. The parallels that can be drawn between the phenotypic properties of the DATKO mice and certain symptoms and drug responses of individuals with ADHD raise the possibility that common mechanisms might underlie the pharmacological actions of psychostimulants if not their phenotypes (Gainetdinov et al. 1999b; Gainetdinov and Caron 2000, 2001).

Evidence also suggests that, in the DAT-KO mice, an involvement of glutamatergic neurotransmission in the interplay between the dopaminergic and serotonergic systems can be revealed. Treatment of DATKO mice with the NMDA receptor antagonist MK801 further increases their enhanced locomotor activity and antagonizes the calming effects of psychostimulants and serotonergic drugs (Gainetdinov et al. 2001a). These results raise the possibility that the serotonergic modulation of the hyperactivity behavior in DATKO mice may be through regulation of the glutamatergic input to the striatal complex controlling locomotion (Fig. 1). Elucidating which of the 14 distinct subtypes of serotonin receptors (Kroeze and Roth 1998) modulate the glutamatergic activity of the fronto-cortical pathway may provide insights into novel therapeutic opportunities for ADHD.

Overall, the DAT-KO mice recapitulate several behavioral features of ADHD. These include hyperactivity, impairments in cognitive tasks and a paradoxical inhibition of locomoting in response to psychostimulants (Gainetdinov et al. 1999b; Gainetdinov and Caron 2000, 2001). Interestingly, mice that harbor a reduced expression of DAT (<10% in the striatum) also demonstrate hyperactivity, impaired response habituation and paradoxical responses to amphetamine (Zhuang et al. 2001). In contrast, transgenic mice that have a slight increase in DAT expression (~ 20–30%) show hypoactivity that is evident in a novel environment (Donovan et al. 1999).

The characterization of animals as a model of human disorders, as stated at the outset of this paper, carries obvious caveats. Even if the DAT gene were involved in ADHD, it is unlikely that individuals with ADHD have a complete functional absence of DAT. As such, the DAT-KO mice represent an extreme case of a potential DAT dysfunction. As in the case with the symptoms of schizophrenia, it should be emphasized that potential dopaminergic dysregulation in ADHD, as that observed in the DAT-KO mice, might be produced by defects in many different components of the system other than DAT. Moreover, dysregulating the DA system is unlikely to be the exclusive mechanism involved in the development of ADHD-like conditions.

## Other genetically altered mice that may have potential implications for understanding schizophrenia

Characterization of mutant mice for behaviors related to schizophrenia has already revealed some interesting phenotypes. It is clear that there are multiple ways to affect critical components of frontostriatal circuitry that may result in the development of aberrant behaviors. Psychotic symptoms may result from alterations in neurotransmitter concentration, receptor number or subunit composition, or perhaps from changes in downstream signaling from receptors. Several neurotransmitter systems have been implicated in the pathophysiology of schizophrenia, but the most prominent systems appear to be DA, glutamate and serotonin (Carlsson et al. 2001). Therefore, we have limited discussion to the phenotypes of mice with targeted mutations in DA, glutamate and serotonin systems only, with emphasis on those behaviors that have been traditionally correlated to symptoms of schizophrenia. The aberrant behaviors primarily considered are increased motor activity and stereotypy, deficits in social behaviors, and sensorimotor gating. Unfortunately the available data are limited in some instances, and additional in-depth studies will be necessary to evaluate these models. In addition, in many cases, the efficacy of antipsychotic drugs in modulating the respective phenotypes has not yet been fully examined.

Knockouts of all of the DA receptors (D1-D5) have been developed (Sibley 1999). One of the surprising findings from these studies is that locomotor activity in a basal, non-challenging condition is not remarkably affected. D2 DA receptor knockout mice demonstrate some hypoactivity (Baik et al. 1995; Kelly et al. 1998), D3 DA receptor mutants show modest hyperactivity (Accili et al. 1996; Xu et al. 1997), and D1 DA receptor mutants are reported to have a moderate hyperactivity (Xu et al. 1994; Clifford et al. 1998), whereas D4 DA receptor mutants seem to show slightly decreased activity in a novel environment (Rubinstein et al. 1997; Dulawa et al. 1999). Unfortunately, some of these observations were not replicated in subsequent studies (Drago et al. 1994; Smith et al. 1998; Jung et al.

1999). Mice have also been generated with mutations in the genes encoding the major enzymes involved in DA synthesis and metabolism. Deletion of the gene encoding a key limiting enzyme in DA biosynthesis, tyrosine hydroxylase, results, as might be predicted, in severely impaired locomotor activity that can be restored with the DA precursor L-DOPA (Zhou and Palmiter 1995; Kim et al. 2000). Genetic deletion of catechol-O-methyltransferase (COMT), the major enzyme involved in extraneuronal metabolic degradation of DA, did not result in any significant alterations in locomotor activity or sensorimotor gating, but impairment in emotional reactivity and increased aggression were noted (Gogos et al. 1998). Mice with a deletion of monoamine oxidase A (MAO-A), a major determinant of intraneuronal monoamine metabolism, have remarkable alterations in serotonin and norepinephrine metabolism with little effect on DA function (Shih et al. 1999). The major phenotype described in these mice is aggressive behavior. Behaviors related to schizophrenia have not been tested. MAO-B knockout mice have elevated levels of only phenylethylamine and do not display aggressive behaviors. Like MAO-A mutants, they are more reactive to stress (Shih et al. 1999). None of the aforementioned mice were reported to have signs of social withdrawal, and only D2 DA receptor knockouts demonstrate some dysfunction in mechanisms of pre-pulse regulation (Ralph et al. 1999). Thus, it is already clear that neither knockouts of DA receptors nor of DA-related enzymes seem likely models for schizophrenia-related behaviors. However, these mutations might be expected to significantly modulate the phenotypes of other mutant mice that may be more closely relevant to schizophrenia.

The observation of schizophrenia-related behaviors in NR1-deficient mice is consistent with a role for the glutamate system in schizophrenia (Mohn et al. 1999). The role of NR1 in schizophrenia-related behaviors has received further support recently with the characterization of mice carrying point mutations in the NMDA receptor glycine binding site, which markedly reduces the affinity of glycine for the NMDA receptor (Ballard et al. 2002). These mutants exhibited a marked NMDA receptor hypofunction revealed by deficits in hippocampal long-term potentiation, which were rescued by the glycine site agonist D-serine. Analysis of striatal monoamine levels suggested an apparent dopaminergic and serotonergic hyperfunction. These mice were insensitive to (+)-MK-801 pretreatment and exhibited an increased startle response. Like NR1-KD mice (Mohn et al. 1999), mutant mice exhibited hyperactivity and increased stereotyped behavior. However, low doses of antipsychotics and the benzodiazepine site agonist Zolpidem showed only minor effects on this hyperactivity. Interestingly, a disruption of nest building behavior and inability to perform a cued learning paradigm in the Morris water maze were also noted (Ballard et al. 2002).

To date, mice with a genetic deletion of all of the major subunits of NMDA and AMPA glutamate receptors have been developed and characterized. Global deletion of NR1 (Forrest et al. 1994; Li et al. 1994) and NR2B receptor (Kutsuwada et al. 1996) subunits of the NMDA receptor complex results in perinatal lethality that prevents evaluation of the role of these molecules in psychotic-related behaviors. Among the other glutamatergic mutants, the mice lacking the NR2A subunit of NMDA receptor (Sakimura et al. 1995; Kadotani et al. 1996; Miyamoto et al. 2001) are of particular interest. NR2A mutant mice exhibited an

increased locomotor activity in a novel environment and cognitive deficits manifested in various learning tasks (Sakimura et al. 1995). Hyperlocomotion in
NR2A mutant mice was attenuated by treatment with haloperidol and risperi-
done (Miyamoto et al. 2001). In a preliminary neurochemical investigation, both
DA and serotonin metabolism were found to be modestly affected in the frontal
cortex and striatum of these mutant mice. In addition, the NMDA-stimulated
$[^3H]$dopamine release from striatal slices was increased, whereas $[^3H]$GABA release was markedly diminished (Miyamoto et al. 2001). NR2D knockout mice
develop normally but show some impairment in learning and motor coordination (Ikeda et al. 1995). An increase in dendritic spine density and no obvious
behavioral abnormalities were found in NR3A subunit knockout mice (Das et al.
1998). None of these mutant lines were reported to display hyperactivity or signs
of social withdrawal. Mice lacking the GluR1 subunit of AMPA receptors showed
an absence of long-term potentiation in the hippocampal region, surprisingly
without significant impairment in their ability to learn (Zamanillo et al. 1999).
GluR2 mutant mice showed enhanced LTP, reduced exploration, and impaired
motor coordination (Jia et al. 1996).

Metabotropic glutamate receptor mutants (Aiba et al. 1994; Masu et al. 1995;
Pekhletski et al. 1996; Lu et al. 1997; Masugi et al. 1999; Spooren et al. 2000) may
also be attractive models for investigating schizophrenia-related behaviors
(Moghaddam and Adams 1998); however, a potential role for these molecules in
aberrant behaviors remains unclear.

Mutations in several components of glutamatergic transmission could result
in schizophrenia-related phenotypes akin to those seen in NR1-KD. Deciphering
the components of the glutamatergic neuronal system that are involved in the
development of this phenotype represents a major challenge for future research.
In addition to the NMDA, AMPA and metabotropic glutamate receptors, there
are many other genes that could contribute to these phenotypes. These include
kinases and phosphatases that modulate the activity of receptors, proteins involved in synthesis, metabolism and uptake of glutamate and the co-agonists
glycine and D-serine, glutamate-related peptides, and proteins involved in the
trafficking of AMPA and metabotropic glutamate receptors, etc.

The role of serotonergic neurotransmission in psychotic behaviors is less well
understood. Nevertheless, it is already clear that delineation of the functions of
serotonin receptors is extremely important for understanding the pathogenesis
of schizophrenia and the mechanism of action of antipsychotic drugs (Martin et
al. 1998; Meltzer et al. 1999; Aghajanian and Marek 2000). For example, it is
known that several serotonin receptors (including 5-HT2A and 5-HT1B receptors) can exert a stimulatory action on behavior, whereas others are inhibitory
(primary candidates are 5-HT1A and 5-HT2C receptors; Martin et al. 1998). Re-
storing the balance between these opposing actions of serotonin has been recently suggested as a primary determinant of the antipsychotic action of atypical
neuroleptics (Martin et al. 1998). To date, at least 14 serotonin receptors have
been identified by molecular cloning. Most of the pharmacological agents available demonstrate only limited selectivity for these receptors, and for several of
them no selective agents are yet available (Kroeze and Roth 1998). At present, an

investigation of mutant mice may represent the only available means of understanding the function of each receptor (Murphy et al. 1999).

Mice lacking 5-HT1A, 5-HT1B, 5-HT2C and 5-HT5A have been generated (Zhuang et al. 1999; Heisler and Tecott 2000; Grailhe et al.,1999). 5HT1A mutant mice show increased anxiety and increased sensitivity to antidepressants; 5-HT1B knockout mice show hyperactivity, aggression and numerous alterations in tests of emotional states (Zhuang et al. 1999). Interestingly, both 5-HT1A and 5-HT1B mutant mice show significant alterations in the regulation of sensorimotor gating (Dulawa et al. 2000). The basal level of activity was not altered in 5HT2C knockout mice. However, the effects of several drugs on locomotion were blunted, suggesting an inhibitory influence of 5-HT2C receptors on motor activity (Heisler and Tecott 2000). 5-HT5A knockout mice showed a decreased rate of locomotor habituation to novel environment but no alteration in sensorimotor gating (Grailhe et al. 1999). 5-HT transporter knockout mice showed no alterations in basal level of locomotor activity; however, locomotor response to 3,4-methylenedioxymethamphetamine (MDMA) was blocked (Bengel et al. 1998). At present there is no available information on mice bearing mutations in the genes for 5-HT2A, 5-HT6 and 5-HT7 serotonin receptors or for the critical enzyme in 5-HT synthesis, tryptophan hydroxylase. It is anticipated that further characterization of serotonergic systems will provide important information on the etiology and treatment principles of schizophrenia.

The other neurotransmitter systems, such as cholinergic (particularly, muscarinic), that are potently involved in the modulation of various components of frontostriatal circuitry (Holt et al. 1999) also represent attractive targets for targeted mutagenesis. Recently mice lacking the M1 muscarinic acetylcholine receptor were generated and examined for the effects of M1 deletion on dopaminergic transmission and locomotor behavior (Gerber et al. 2001). An elevated dopaminergic transmission in the striatum and significantly increased locomotor activity were observed in these mice. M1-deficient mice also have an increased response to the stimulatory effects of amphetamine. Thus, direct evidence for regulation of dopaminergic transmission by the M1 receptor was demonstrated. Accordingly, it is possible that M1 dysfunction could be a contributing factor in psychiatric disorders in which altered dopaminergic transmission has been implicated (Gerber et al. 2001). It might also be expected that mutations in the noradrenergic system would result in aberrant behaviors relevant to schizophrenia (Hertel et al. 1999). However, the insufficient status of current knowledge prevents a definitive prognosis as to which of the components of these systems would be the most likely candidates.

Mutations in other genes may result in a phenotype that can be relevant to schizophrenia. For example, mice lacking DVL1 (a homolog of the Drosophila segment polarity gene *disheveled*) demonstrate reduced social interaction and deficits in PPI of acoustic and tactile startle response (Lijam et al. 1997). Mice lacking the neural cell adhesion molecule (NCAM-180) also display deficit in PPI as well as increased lateral ventricle size (Wood et al. 1998). Altered psychomotor behaviors were described in mice lacking pituitary adenylate cyclase-activating polypeptide (PACAP; Hashimoto et al. 2001). However, a better understanding of the role of such proteins, particularly in the human brain, is required to validate their potential role in schizophrenia. A recent exciting report on the role of neuregulin 1 (NRG1) in schizophrenia may represent an example of such validation (Stefansson et al. 2002). A genome-wide scan of schizophrenia families in Iceland showed that schizophrenia maps to chromosome 8p. Extensive fine-mapping of the chromosome 8p locus and haplotype-association analysis identified NRG1 as a candidate gene for schizophrenia. NRG1 is expressed at CNS synapses and has an important role in the regulation of neurotransmitter receptors, including glutamate receptors. Mutant mice heterozygous for either NRG1 or its receptor, ErbB4, showed fewer functional NMDA receptors and a behavioral phenotype similar to NR1-KD mice. Furthermore, the behavioral phenotypes of the NRG1 hypomorphs were partially reversible with clozapine, an atypical antipsychotic drug used to treat schizophrenia (Stefansson et al. 2002).

## Potential implications of genetic models

It is already clear from the studies on DAT-KO and NR1-KD mice that these mice support many of the findings discovered previously by pharmacological approaches. In particular, genetic models have validated the interaction of DA and glutamate systems in the control of locomotion that was proposed based on the effects of pharmacological agents (Carlsson et al. 2001). The ability to test various hypotheses of schizophrenia using genetic approaches will open the door to more precisely clarifying how neurotransmitter systems interact to develop psychotic behaviors and to understanding the consequential order of events involved in these manifestations. For example, the complexity of glutamateDA-interactions in the control of locomotion provided the basis for a surge of hypotheses about how these systems interact. Many of these theoretical constructs contradict each other and verification of the salient properties of these hypotheses with genetic models could provide important clarifications of several critical points for understanding schizophrenia.

The simplified diagram presented here (Fig. 1), which encompasses common features of several previous models (Javitt and Zukin 1991; Carlsson et al. 2001), is appropriate for explaining most aberrant behaviors of genetically altered animals (Mohn et al. 1999; Gainetdinov et al. 2001a). This reciprocal interaction between DA and glutamate systems strongly suggests that locomotor activity, considered here as an index of aberrant behavior, can be influenced by glutamatergic drugs (Coyle 1996; Johnson et al. 1999; Carlsson et al. 2001). In particular, it suggests that glutamatergic deficiency could exacerbate aberrant behaviors induced by enhanced dopaminergic tone. In addition, it implies that drugs that increase glutamatergic transmission could counteract dopaminergic hyperactivity. In the search for such agents, two major targets are currently receiving attention. One target is the AMPA receptor, and drugs that inhibit desensitization of AMPA receptors (AMPAkines) may increase efficiency of glutamatergic transmission by modulating NMDA receptor currents (Johnson et al. 1999; Gainetdinov et al. 2001a). Alternatively, drugs that affect the glycine-binding site on the NMDA receptor (glycinergics) may also be useful in enhancing NMDA receptor function (Coyle 1996; Klitenick et al. 2001). Both avenues of research have yielded tantalizing results in several pre-clinical pharmacological studies; however, more studies are required to validate their potential utility in treating schizophrenia.

With regards to serotonergic transmission, modulation of glutamate neuron activity in the frontal cortex by serotonin is the most attractive postulate (Martin et al. 1998; Aghajanian and Marek 2000). From attempts to understand the mechanism of action of novel antipsychotics, it has become clear that serotonin, potentially through modulation of glutamatergic transmission, can exert a potent modulatory effect on behavior. However, these effects may not be simple, as different serotonin receptor subtypes may exert either excitatory or inhibitory actions on behavior (Martin et al. 1998). These observations suggest that antipsychotic action may be obtained by developing drugs that act selectively on specific serotonin receptor subtypes. Future studies employing crosses between animals with specific alterations in components of DA, glutamate and serotonin systems could be helpful in defining the role of each of the serotonin receptors in the modulation of frontostriatal circuitry and, eventually, aberrant behaviors.

One benefit of using genetically altered animals as models of human diseases is that these mice could become excellent models for future drug development. These models provide pharmacologists a long-awaited opportunity to test potential therapeutic agents in animals with an well-studied brain pathology resulting from a specific genetic manipulation. It might be expected that different classes of drugs will show different efficacies in different mutant animals. As an example, different profiles of activity of typical and atypical neuroleptics have already been observed in hyperdopaminergic or hypoglutamatergic models of brain pathology (Spielewoy et al. 2000; Mohn et al. 1999). These studies may lead the way to a better prognosis of the efficacy of novel antipsychotics in various subtypes of schizophrenia. The future use of these animals may include not only their utility in screening of new antipsychotic drugs but also their potential in elucidating the mechanisms by which dopamine or glutamate can modulate the activity of other neurotransmitter systems. Particularly useful in this respect will be the development of mice with region-selective, inducible genetic alterations, where only specific neuronal groups are affected. Genetically altered mice may become increasingly valuable for understanding the contribution of various neurotransmitter systems of the cortico-basal ganglia circuits in manifestation of aberrant behaviors. Using DAT-KO mice as a test example, it should be possible to further clarify the anatomical sites and receptor specificity involved in the inhibitory action of serotonin on dopaminergic hyperactivity (Gainetdinov et al. 1999b). It will be of interest also to determine the impact of chronic NMDA receptor downregulation through development not only on the behavioral manifestations but also on the activity and expression of other neurotransmitter systems, as well as on the formation and elimination of synaptic connections. As many of the CNS disorders are likely to result from the aberrant function of several genes or neurotransmitter systems, perhaps the greatest potential of genetically altered animals is in the production of compound-heterozygote animals. In such animals, partial loss or gain of function in several functional components could recapitulate more precisely the etiology and symptoms of a complex condition such as schizophrenia.

In conclusion, studies on genetically altered animals are relevant to schizophrenia and other neuropsychiatric conditions for several reasons. Mutant mice are valuable tools for dissecting out the contribution of specific neurotransmitter systems within the frontostriatal circuitry. This circuitry is believed to be central to the pathophysiology of several psychiatric disorders. The use of genetically altered mice as models for human disorders will allow testing of the hypotheses about the cause or symptoms of these disorders in humans. Such mice will also be substrates for evaluating the effects of potential antipsychotic drugs in animals with known genetic and biochemical alterations. In addition these studies could identify a role for novel candidate genes for these disorders, through genetic linkage or genome scans. These approaches will utilize the power of genetic manipulations in the whole animal to complement the previously available wealth of behavioral, cellular, and pharmacological information in an attempt to elucidate the principles underlying manifestations of psychotic symptoms.

## Acknowledgments

Work from this laboratory was supported in part by grants from NIH: NS 19576, 1 RO1 DA131511. M.G. Caron is an Investigator of the Howard Hughes Medical Institute. e-mail addresses: MGC: CARON002@mc.duke.edu RRG: R.Gainetdinov@cellbio.duke.edu

## References

Accili D, Fishburn CS, Drago J, Steiner H, Lachowicz JE, Park BH, Gauda EB, Lee EJ, Cool MH, Sibley DR, Gerfen CR, Westphal H, Fuchs S. (1996) A targeted mutation of the D3 dopamine receptor gene is associated with hyperactivity in mice. Proc Natl Acad Sci USA 93: 1945–1949

Aghajanian GK, Marek GJ (2000) Serotonin model of schizophrenia: emerging role of glutamate mechanisms. Brain Res Rev 31: 302–312

Aiba A, Kano M, Chen C, Stanton ME, Fox GD, Herrup K, Zwingman TA, Tonegawa S (1994) Reduced hippocampal long-term potentiation and context-specific deficit in associative learning in mGluR1 mutant mice. Cell 79: 365–375

Baik JH, Picetti R, Saiardi A, Thiriet G, Dierich A, Depaulis A, Le Meur M, Borrelli E. (1995) Parkinsonian-like locomotor impairment in mice lacking dopamine D2 receptors. Nature 377: 424–428

Ballard TM, Pauly-Evers M, Higgins GA, Ouagazzal AM, Mutel V, Borroni E, Kemp JA, Bluethmann H, Kew JN (2002) Severe impairment of NMDA receptor function in mice carrying targeted point mutations in the glycine binding site results in drug-resistant nonhabituating hyperactivity. J Neurosci 22: 6713–6723

Barkley RA (1990) Attention deficit hyperactivity disorder: a handbook for diagnosis and treatment, Guiford Press, New York

Bengel D, Murphy DL, Andrews AM, Wichems CH, Feltner D, Heils A, Mossner R, Westphal H, Lesch KP (1998) Altered brain serotonin homeostasis and locomotor insensitivity to 3, 4-methylenedioxymethamphetamine ("Ecstasy") in serotonin transporter-deficient mice. Mol Pharmacol 53: 649–655

Bodeau-Pean S, Laurent C, Campion D, Jay M, Thibaut F, Dollfus S, Petit M, Samolyk D, D'Amato T, Martinez M, Mallet J (1995) No evidence for linkage or association between the dopamine transporter gene and schizophrenia in a French population. Psychiat Res 59 : 1–6

Bosse R, Fumagalli F, Jaber M, Giros B, Gainetdinov RR, Wetsel WC, Missale C, Caron MG (1997) Anterior pituitary hypoplasia and dwarfism in mice lacking the dopamine transporter. Neuron 19: 127–138

Boulay D, Depoortere R, Oblin A, Sanger DJ, Schoemaker H, Perrault G (2000) Haloperidol-induced catalepsy is absent in dopamine D(2), but maintained in dopamine D(3) receptor knock-out mice. Eur J Pharmacol 391: 63–73

Carlsson A, Waters N, Holm-Waters S, Tedroff J, Nilsson M, Carlsson ML (2001) Interaction between monoamines, glutamate, and GABA in schizophrenia: new evidence. Annu Rev Pharmacol Toxicol 41: 237–260

Clifford JJ, Tighe O, Croke DT, Sibley DR, Drago J, Waddington JL (1998) Topographical evaluation of the phenotype of spontaneous behaviour in mice with targeted gene deletion of the D1A dopamine receptor: paradoxical elevation of grooming syntax. Neuropharmacology 37 : 159561602

Cook EH Jr, Stein MA, Krasowski MD, Cox NJ, Olkon DM, Kieffer JE, Leventhal BL (1995) Association of attention-deficit disorder and the dopamine transporter gene. Am J Human Genet 56: 993–998

Corbett R, Camacho F, Woods AT, Kerman LL, Fishkin RJ, Brooks K, Dunn RW (1995) Antipsychotic agents antagonize non-competitive N-methyl-D-aspartate antagonist-induced behaviors. Psychopharmacology (Berl) 120: 67–74

Coyle JT (1996) The glutamatergic dysfunction hypothesis for schizophrenia. Harv Rev Psychiat 3: 241–253

Creese I, Burt DR, Snyder SH (1976) Dopamine receptor binding predicts clinical and pharmacological potencies of antischizophrenic drugs. Science 192: 481–483

Das S, Sasaki YF, Rothe T, Premkumar LS, Takasu M, Crandall JE, Dikkes P, Conner DA, Rayudu PV, Cheung W, Chen HS, Lipton SA, Nakanishi N (1998) Increased NMDA current and spine density in mice lacking the NMDA receptor subunit NR3A. Nature 393: 377–381

Donovan DM, Miner LL, Perry MP, Revay RS, Sharpe LG, Przedborski S, Kostic V, Philpot RM, Kirstein CL, Rothman RB, Schindler CW, Uhl GR (1999) Cocaine reward and MPTP toxicity: alteration by regional variant dopamine transporter overexpression. Mol Brain Res 73: 37–49

Drago J, Gerfen CR, Lachowicz JE, Steiner H, Hollon TR, Love PE, Ooi GT, Grinberg A, Lee EJ, Huang SP, Bartlett PF, Jose PA, Sibley DR, Westphal H (1994) Altered striatal function in a mutant mouse lacking D1A dopamine receptors. Proc Natl Acad Sci USA 91: 12564–12568

Dulawa SC, Grandy DK, Low MJ, Paulus MP, Geyer MA (1999) Dopamine D4 receptor-knock-out mice exhibit reduced exploration of novel stimuli. J Neurosci 19: 9550–9556

Dulawa SC, Gross C, Stark KL, Hen R, Geyer MA (2000) Knockout mice reveal opposite roles for serotonin 1A and 1B receptors in prepulse inhibition. Neuropsychopharmacology 22: 650–659

El-Ghundi M, Fletcher PJ, Drago J, Sibley DR, O'Dowd BF, George SR (1999) Spatial learning deficit in dopamine D(1) receptor knockout mice. Eur J Pharmacol 383: 95–106

Ellinwood EH Jr (1967) Amphetamine psychosis: I. Description of the individuals and process. J Nerv Ment Dis 144: 273–283

Fauchey V, Jaber M, Caron MG, Bloch B, Le Moine C (2000) Differential regulation of the dopamine D1, D2 and D3 receptor gene expression and changes in the phenotype of the striatal neurons in mice lacking the dopamine transporter. Eur J Neurosci 12: 19–26

Forrest D, Yuzaki M, Soares HD, Ng L, Luk DC, Sheng M, Stewart CL, Morgan JI, Connor JA, Curran T (1994) Targeted disruption of NMDA receptor 1 gene abolishes NMDA response and results in neonatal death. Neuron 13: 325–338

Gainetdinov RR, Caron MG (2000) An animal model of attention deficit hyperactivity disorder. Mol Med Today 6: 43–44

Gainetdinov RR, Caron MG (2001) Genetics of childhood disorders: XXIV. ADHD, part 8: Hyperdopaminergic mice as an animal model of ADHD. J Am Acad Child Psychiat 40: 380–382

Gainetdinov RR, Jones SR, Fumagalli F, Wightman RM, Caron MG (1998) Re-evaluation of the role of the dopamine transporter in dopamine system homeostasis. Brain Res Rev 26: 148–153

Gainetdinov RR, Jones SR, Caron MG (1999a) Functional hyperdopaminergia in dopamine transporter knock-out mice. Biol Psychiat 46: 303–311

Gainetdinov RR, Wetsel WC, Jones SR, Levin ED, Jaber M, Caron MG (1999b) Role of serotonin in the paradoxical calming effect of psychostimulants on hyperactivity. Science 283: 397–401

Gainetdinov RR, Mohn AR, Bohn LM, Caron MG (2001a) Glutamatergic modulation of hyperactivity in mice lacking the dopamine transporter. Proc Natl Acad Sci USA 98: 11047–11054

Gainetdinov RR, Mohn AR, Caron MG (2001b) Genetic animal models: focus on schizophrenia. Trends Neurosci 24: 527–533

Gainetdinov RR, Sotnikova TD, Caron MG (2002) Monoamine transporter pharmacology and mutant mice. Trends Pharmacol Sci 23: 367–374.

Gerber DJ, Sotnikova TD, Gainetdinov RR, Huang SY, Caron MG, Tonegawa S (2001) Hyperactivity, elevated dopaminergic transmission, and response to amphetamine in M1 muscarinic acetylcholine receptor-deficient mice. Proc Natl Acad Sci USA 98: 15312–15317

Gill M, Daly G, Heron S, Hawi Z, Fitzgerald M (1997) Confirmation of association between attention deficit hyperactivity disorder and a dopamine transporter polymorphism. Mol Psychiatr 2: 311–13

Giros B, Jaber M, Jones SR, Wightman RM, Caron MG (1996) Hyperlocomotion and indifference to cocaine and amphetamine in mice lacking the dopamine transporter. Nature 379: 606–612

Gogos JA, Morgan M, Luine V, Santha M, Ogawa S, Pfaff D, Karayiorgou M (1998) Catechol-O-methyltransferase-deficient mice exhibit sexually dimorphic changes in catecholamine levels and behavior. Proc Natl Acad Sci USA 95: 9991–9996

Grailhe R, Waeber C, Dulawa SC, Hornung JP, Zhuang X, Brunner D, Geyer MA, Hen R (1999) Increased exploratory activity and altered response to LSD in mice lacking the 5-HT(5A) receptor. Neuron 22: 581–591

Greengard P (2001) The neurobiology of slow synaptic transmission. Science 294: 1024–1030

Hashimoto H, Shintani N, Tanaka K, Mori W, Hirose M, Matsuda T, Sakaue M, Miyazaki J, Niwa H, Tashiro F, Yamamoto K, Koga K, Tomimoto S, Kunugi A, Suetake A S, Baba (2001) Altered psychomotor behaviors in mice lacking pituitary adenylate cyclase-activating polypeptide (PACAP). Proc Natl Acad Sci USA 98: 13355–13360

Heisler LK, Tecott LH (2000) A paradoxical locomotor response in serotonin 5-HT(2C) receptor mutant mice. J Neurosci 20: RC71

Hertel P, Fagerquist MV, Svensson TH (1999) Enhanced cortical dopamine output and antipsychotic-like effects of raclopride by alpha2 adrenoceptor blockade. Science 286: 105–107

Holt DJ, Herman MM, Hyde TM, Kleinman JE, Sinton CM, German DC, Hersh LB, Graybiel AM, Saper CB (1999) Evidence for a deficit in cholinergic interneurons in the striatum in schizophrenia. Neuroscience 94 : 21–31

Ikeda K, Araki K, Takayama C, Inoue Y, Yagi T, Aizawa S, Mishina M (1995) Reduced spontaneous activity of mice defective in the epsilon 4 subunit of the NMDA receptor channel. Mol Brain Res 33: 61–71

Inada T, Sugita T, Dobashi I, Inagaki A, Kitao Y, Matsuda G, Kato S, Takano T, Yagi G, Asai M (1996) Dopamine transporter gene polymorphism and psychiatric symptoms seen in schizophrenic patients at their first episode. Am J Med Genet 67: 406–408

Javitt DC, Zukin SR (1991) Recent advances in the phencyclidine model of schizophrenia. Am J Psychiat 148: 1301–1308

Jia Z, Agopyan N, Miu P, Xiong Z, Henderson J, Gerlai R, Taverna FA, Velumian A, MacDonald J, Carlen P, Abramow-Newerly W, Roder J (1996) Enhanced LTP in mice deficient in the AMPA receptor GluR2. Neuron 17: 945–956

Johnson SA, Luu NT, Herbst TA, Knapp R, Lutz D, Arai A, Rogers GA, Lynch G (1999) Synergistic interactions between ampakines and antipsychotic drugs. J Pharmacol Exp Ther 289: 392–397

Jones SR, Gainetdinov RR, Jaber M, Giros B, Wightman RM, Caron MG (1998a) Profound neuronal plasticity in response to inactivation of the dopamine transporter. Proc Natl Acad Sci USA 95: 4029–4034

Jones SR, Gainetdinov RR, Wightman RM, Caron MG (1998b) Mechanisms of amphetamine action revealed in mice lacking the dopamine transporter. J Neurosci 18: 1979–1986

Jones SR, Gainetdinov RR, Hu X-T, Cooper DC, Wightman RM, White FJ, Caron MG (1999) Loss of autoreceptor functions in mice lacking the dopamine transporter. Nature Neurosci 2: 649–655

Jung MY, Skryabin BV, Arai M, Abbondanzo S, Fu D, Brosius J, Robakis NK, Polites HG, Pintar JE, Schmauss C (1999) Potentiation of the D2 mutant motor phenotype in mice lacking dopamine D2 and D3 receptors. Neuroscience 91: 911–924

Kadotani H, Hirano T, Masugi M, Nakamura K, Nakao K, Katsuki M, Nakanishi S (1996) Motor discoordination results from combined gene disruption of the NMDA receptor NR2A and NR2C subunits, but not from single disruption of the NR2A or NR2C subunit. J Neurosci 16: 7859–7867

Kelly MA, Rubinstein M, Phillips TJ, Lessov CN, Burkhart-Kasch S, Zhang G, Bunzow JR, Fang Y, Gerhardt GA, Grandy DK, Low MJ (1998) Locomotor activity in D2 dopamine receptor-deficient mice is determined by gene dosage, genetic background, and developmental adaptations. J Neurosci 18: 3470–3479

Kim DS, Szczypka MS, Palmiter RD (2000) Dopamine-deficient mice are hypersensitive to dopamine receptor agonists. J Neurosci 20: 4405–4413

Klitenick MA, Atkinson BN, Baker DA, Bakker M, Bell SC (2001) Development and characterization of GLYT1-selective glycine reuptake inhibitors. In: Abstracts of 40th Annual ACNP meeting, Waikoloa, Hawaii, p. 271

Kokkinidis L, Anisman H (1981) Amphetamine psychosis and schizophrenia: a dual model. Neurosci Biobehav Rev 5: 449–461

Kroeze WK, Roth BL (1998) The molecular biology of serotonin receptors: therapeutic implications for the interface of mood and psychosis. Biol Psychiat 44: 1128–1142

Kutsuwada T, Sakimura K, Manabe T, Takayama C, Katakura N, Kushiya E, Natsume R, Watanabe M, Inoue Y, Yagi T, Aizawa S, Arakawa M, Takahashi T, Nakamura Y, Mori H, Mishina M (1996) Impairment of suckling response, trigeminal neuronal pattern formation, and hippocampal LTD in NMDA receptor epsilon 2 subunit mutant mice. Neuron 16: 333–344

Laruelle M, Abi-Dargham A, Gil R, Kegeles L, Innis R (1999) Increased dopamine transmission in schizophrenia: relationship to illness phases. Biol Psychiat 46: 56–72

Lehmann-Masten VD, Geyer MA (1991) Spatial and temporal patterning distinguishes the locomotor activating effects of dizocilpine and phencyclidine in rats. Neuropharmacology 30: 629–636

Lewis DA, Lieberman JA (2000) Catching up on schizophrenia. Natural history and neurobiology. Neuron 28: 325–334

Li Y, Erzurumlu RS, Chen C, Jhaveri S, Tonegawa S (1994) Whisker-related neuronal patterns fail to develop in the trigeminal brainstem nuclei of NMDAR1 knockout mice. Cell 76: 427–437

Lijam N, Paylor R, McDonald MP, Crawley JN, Deng CX, Herrup K, Stevens KE, Maccaferri G, McBain CJ, Sussman DJ, Wynshaw-Boris A (1997) Social interaction and sensorimotor gating abnormalities in mice lacking Dvl11. Cell 90: 895–905

Lu YM, Jia Z, Janus C, Henderson JT, Gerlai R, Wojtowicz JM, Roder JC (1997) Mice lacking metabotropic glutamate receptor 5 show impaired learning and reduced CA1 long-term potentiation (LTP) but normal CA3 LTP. J Neurosci 17: 5196–5205

Luby ED, Cohen BD, Rosenbaum F, Gottlieb J, Kelley R (1959) Study of a new schizophrenomimetic drug. Sernyl Arch Neurol Psychiat 81: 363–369

Martin P, Waters N, Schmidt CJ, Carlsson A, Carlsson ML (1998) Rodent data and general hypothesis: antipsychotic action exerted through 5-HT2A receptor antagonism is dependent on increased serotonergic tone. J Neural Transm 105: 365–396

Masu M, Iwakabe H, Tagawa Y, Miyoshi T, Yamashita M, Fukuda H, Sasaki H, Hiroi K, Nakamura Y, Shigemoto R (1995) Specific deficit of the ON response in visual transmission by targeted disruption of the mGluR6 gene. Cell 80: 757–765

Masugi M, Yokoi M, Shigemoto R, Muguruma K, Watanabe Y, Sansig G, van der Putten H, Nakanishi S (1999) Metabotropic glutamate receptor subtype 7 ablation causes deficit in fear response and conditioned taste aversion. J Neurosc. 19: 955–963

Meltzer HY (1999) The role of serotonin in antipsychotic drug action. Neuropsychopharmacology 21: 106S–115S

Miyamoto Y, Yamada K, Noda Y, Mori H, Mishina M, Nabeshima T (2001) Hyperfunction of dopaminergic and serotonergic neuronal systems in mice lacking the NMDA receptor epsilon 1 subunit. J Neurosci 21: 750–757

Moghaddam B, Adams BW (1998) Reversal of phencyclidine effects by a group II metabotropic glutamate receptor agonist in rats. Science 281: 1349–1352

Mohn AR, Gainetdinov RR, Caron MG, Kohler BH (1999) Mice with reduced NMDA receptor expression display behaviors related to schizophrenia. Cell 98: 427–436

Murphy DL, Wichems C, Li Q, Heils A (1999) Molecular manipulations as tools for enhancing our understanding of 5-HT neurotransmission. Trends Pharmacol Sci 20: 246–252

Pekhletski R, Gerlai R, Overstreet LS, Huang XP, Agopyan N, Slater NT, Abramow-Newerly W, Roder JC, Hampson DR (1996) Impaired cerebellar synaptic plasticity and motor performance in mice lacking the mGluR4 subtype of metabotropic glutamate receptor. J Neurosci 16: 6364–6373

Persico AM, Macciardi F (1997) Genotypic association between dopamine transporter gene polymorphisms and schizophrenia. Am J Med Genet 74: 53–57

Ralph RJ, Varty GB, Kelly MA, Wang YM, Caron MG, Rubinstein M, Grandy DK, Low MJ, Geyer MA (1999) The dopamine D2, but not D3 or D4, receptor subtype is essential for the disruption of prepulse inhibition produced by amphetamine in mice. J Neurosci 19: 4627–4633

Ralph RJ, Paulus MP, Fumagalli F, Caron MG, Geyer MA (2001) Prepulse inhibition deficits and perseverative motor patterns in dopamine transporter knock-out mice: differential effects of D1 and D2 receptor antagonists. J Neurosci 21: 305–313

Rocha BA, Fumagalli F, Gainetdinov RR, Jones SR, Ator R, Giros B, Miller GW, Caron MG (1998) Cocaine self-administration in dopamine transporter knockout mice. Nature Neurosci 1: 132–137

Rodriguiz RM, Chu R, Stout RD, Pogorelov VM, Caron MG, Wetsel WC (2001) Impulsivity and inattention are features of an ADHD-like phenotype in DAT-KO mice. Abstr Soc Neurosci 27: 539.10.

Rubinstein M, Phillips TJ, Bunzow JR, Falzone TL, Dziewczapolski G, Zhang G, Fang Y, Larson JL, McDougall JA, Chester JA, Saez C, Pugsley TA, Gershanik O, Low MJ, Grandy DK (1997) Mice lacking dopamine D4 receptors are supersensitive to ethanol, cocaine, and methamphetamine. Cell 90: 991–1001.

Sakimura K, Kutsuwada T, Ito I, Manabe T, Takayama C, Kushiya E, Yagi T, Aizawa S, Inoue Y, Sugiyama H, Mishina M (1995) Reduced hippocampal LTP and spatial learning in mice lacking NMDA receptor epsilon 1 subunit. Nature 373: 151–155

Seeman P (1987) Dopamine receptors and the dopamine hypothesis of schizophrenia. Synapse 1: 133–152

Seiden LS, Sabol KE, Ricaurte GA (1993) Amphetamine: effects on catecholamine systems and behavior. Annu Rev Pharmacol Toxicol 33: 639–677

Shih JC, Chen K, Ridd MJ (1999) Monoamine oxidase: from genes to behavior. Annu Rev Neurosci 22: 197–217

Sibley DR (1999) New insights into dopaminergic receptor function using antisense and genetically altered animals. Annu Rev Pharmacol Toxicol 39: 313–341

Smith DR, Striplin CD, Geller AM, Mailman RB, Drago J, Lawler CP, Gallagher M (1998) Behavioural assessment of mice lacking D1A dopamine receptors. Neuroscience 86: 135–146

Smithies O (1993) Animal models of human genetic diseases. Trends Genet 9: 112–116

Snyder SH (1973) Amphetamine psychosis: a "model" schizophrenia mediated by catecholamines. Am J Psychiat 130: 61–67

Sora I, Wichems C, Takahashi N, Li XF, Zeng Z, Revay R, Lesch KP, Murphy DL, Uhl GR (1998) Cocaine reward models: conditioned place preference can be established in dopamine- and in serotonin-transporter knockout mice. Proc Natl Acad Sci USA 95: 7699–7704

Spielewoy C, Roubert C, Hamon M, Nosten-Bertrand M, Betancur C, Giros B (2000) Behavioural disturbances associated with hyperdopaminergia in dopamine-transporter knockout mice. Behav Pharmacol 11: 279–290

Spielewoy C, Biala G, Roubert C, Hamon M, Betancur C, Giros B (2001) Hypolocomotor effects of acute and daily d-amphetamine in mice lacking the dopamine transporter Psychopharmacology (Berl) 159: 2–9

Spooren WP, Gasparini F, van der Putten H, Koller M, Nakanishi S, Kuhn R (2000) Lack of effect of LY314582 (a group 2 metabotropic glutamate receptor agonist) on phencyclidine-induced locomotor activity in metabotropic glutamate receptor 2 knockout mice. Eur J Pharmacol 397: R1–2

Stefansson H, Sigurdsson E, Steinthorsdottir V, Bjornsdottir S, Sigmundsson T, Ghosh S, Brynjolfsson J, Gunnarsdottir S, Ivarsson O, Chou TT, Hjaltason O, Birgisdottir B, Jonsson H, Gudnadottir VG, Gudmundsdottir E, Bjornsson A, Ingvarsson B, Ingason A, Sigfusson S, Hardardottir H, Harvey RP, Lai D, Zhou M, Brunner D, Mutel V, Gonzalo A, Lemke G, Sainz J, Johannesson G, Andresson T, Gudbjartsson D, Manolescu A, Frigge ML, Gurney ME, Kong A, Gulcher JR, Petursson H, Stefansson K (2002) Neuregulin 1 and Susceptibility to Schizophrenia. Am J HumanGenet 71: 877–892

Sulzer D, Chen TK, Lau YY, Kristensen H, Rayport S, Ewing A (1995) Amphetamine redistributes dopamine from synaptic vesicles to the cytosol and promotes reverse transport. J Neurosci 15: 4102–4108

Swerdlow NR, Geyer MA (1998) Using an animal model of deficient sensorimotor gating to study the pathophysiology and new treatments of schizophrenia. Schizophren Bull 24: 285–301

Tamminga CA (1998) Schizophrenia and glutamatergic transmission. Crit Rev Neurobiol 12: 21–36

Waldman ID, Rowe DC, Abramowitz A, Kozel ST, Mohr JH, Sherman SL, Cleveland HH, Sanders ML, Gard JM, Stever C (1998) Association and Linkage of the Dopamine Transporter Gene and Attention- Deficit Hyperactivity Disorder in Children: Heterogeneity owing to Diagnostic Subtype and Severity. Am J Hum Genet 63: 1767–1776

Wood GK, Tomasiewicz H, Rutishauser U, Magnuson T, Quirion R, Rochford J, Srivastava LK (1998) NCAM-180 knockout mice display increased lateral ventricle size and reduced prepulse inhibition of startle. Neuroreport 9: 461–466

Xu M, Moratalla R, Gold LH, Hiroi N, Koob GF (1994) Dopamine D1 receptor mutant mice are deficient in striatal expression of dynorphin and in dopamine-mediated behavioral responses. Cell 79: 729–742

Xu M, Koeltzow TE, Santiago GT, Moratalla R, Cooper DC, Hu XT, White NM, Graybiel AM, White FJ, Tonegawa S (1997) Dopamine D3 receptor mutant mice exhibit increased behavioral sensitivity to concurrent stimulation of D1 and D2 receptors. Neuron 19: 837–848

Zamanillo D, Sprengel R, Hvalby O, Jensen V, Burnashev N, Rozov A, Kaiser KM, Koster HJ, Borchardt T, Worley P, Lubke J, Frotscher M, Kelly PH, Sommer B, Andersen P, Seeburg PH, Sakmann B (1999) Importance of AMPA receptors for hippocampal synaptic plasticity but not for spatial learning. Science 284: 1805-1811

Zhou QY, Palmiter RD (1995) Dopamine-deficient mice are severely hypoactive, adipsic, and aphagic. Cell 83: 1197–1209

Zhuang X, Gross C, Santarelli L, Compan V, Trillat AC, Hen R (1999) Altered emotional states in knockout mice lacking 5-HT1A or 5-HT1B receptors. Neuropsychopharmacology 21(2 Suppl): 52S–60S

Zhuang X, Oosting RS, Jones SR, Gainetdinov RR, Miller GW, Caron MG, Hen R (2001) Hyperactivity and impaired response habituation in hyperdopaminergic mice. Proc Natl Acad Sci USA 98: 1982-87.

# Neurodegenerative Diseases: Insights from *Drosophila* and Mouse Models

*Juan Botas*[1]

## Introduction

*Drosophila melanogaster* provides a flexible and powerful model to study neurodegenerative diseases. Distinct advantages of *Drosophila* as an experimental organism include its small genome, short generation time, and the feasibility of maintaining large numbers of animals in the laboratory, facilitating large-scale mutagenesis screens to isolate genes relevant to any phenotype of interest. *Drosophila* has been used extensively to investigate normal development and its genome has been sequenced (Adams et al. 2000), so that there is a substantial amount of information concerning the conservation of both individual genes and gene pathways between flies and humans (Zipursky and Rubin 1994; Yin and Tully 1996; Shubin et al. 1997). In this context, Warrick et al. (1998) used the fly to study disease progression in a class of human neurodegenerative disorders caused by expansion of polyglutamine repeats within the coding region of the corresponding disease-causing protein (Cummings and Zoghbi 2000). While a normal repeat length has no pathological consequence, expansion of the glutamine tract beyond a critical threshold is associated with neurodegeneration. Truncated fragments of both the normal and mutant forms of human ataxin-3 implicated in spinal cerebellar ataxia type-3 (*SCA3*) were directed to the fly's eye using the GAL4-UAS system for transgene expression (Brand and Perrimon 1993). Targeted expression of the expanded protein led to the generation of nuclear inclusions and late-onset cell degeneration characteristic of SCA3, demonstrating that at least some of the features of cellular pathology in humans are conserved in flies.

Several models of human neurodegenerative disease in *Drosophila* have been reported subsequently in the literature, including those for Huntington's disease (Jackson et al. 1998), Parkinson's disease (Feany and Bender 2000), spinal cerebellar ataxia type-1 (Fernandez-Funez et al. 2000), and the tauopathies (Wittmann et al. 2001). Since these human diseases are not triggered by lack of function of the causative genes, they are particularly well suited for this experimental approach, as reflected in the recapitulation of many of the significant cellular phenotypes in the *Drosophila* models. In addition, *Drosophila* offers the opportunity to identify genetic modifiers that could potentially serve as specific

[1] Department of Molecular and Human Genetics, Baylor College of Medicine, One Baylor Plaza, Houston TX 77030, USA, e-mail: jbotas@bcm.tmc.edu

Mallet/Christen
Neurosciencess at the Postgenomic Era
© Springer-Verlag Berlin Heidelberg 2003

protein targets in future therapeutic interventions. The contributions made to-wards understanding the mechanisms of pathogenesis through the generation of these *Drosophila* disease models will be addressed in this review.

## Polyglutamine Diseases

Nine inherited neurodegenerative diseases caused by the expansion of a CAG repeat tract in the relevant disease protein have been identified to date. These diseases include Huntington's disease (HD), spinobulbar muscular atrophy (SMA), dentatorubral-pallidoluysian atrophy (DRPLA), spinocerebellar ataxia type 1 (SCA1), SCA2, SCA6, SCA7, SCA17, and Machado-Joseph disease (MJD/SCA3). The respective disease-causing proteins seem to have no apparent relationship to each other, and their functions are largely unknown except for those involved in SMBA (the androgen receptor), SCA6 (the a1A calcium channel), DRPLA (atrophin-1, a co-repressor), and SCA17 (the TATA-binding transcription factor; Cummings and Zoghbi 2000; Nakamura et al. 2001; Zhang et al. 2002). As was first demonstrated for SMBA (La Spada et al. 1991), an earlier age of onset and more severe disease have been correlated with an increase in the length of the expanded repeat. Another common but poorly understood feature of the polyglutamine diseases is the neuronal selectivity found for each disease in the presence of widespread gene expression (Reddy and Housman 1997; Ross 1997).

Inactivation of *Hdh*, the mouse HD gene homolog, was undertaken to distinguish between loss of function and gain of function models for the triplet repeat disorders. Mice heterozygous for *Hdh* deficiency were phenotypically normal, whereas homozygous mice displayed abnormal gastrulation and died between embryonic day 8.5 and 10.5 of gestation (Duyao et al. 1995; Zeitlin et al. 1995) so that loss of huntingtin function does not result in disease. Conversely, mouse models for HD, SCA1, and MJD (Burright et al. 1995; Hodgson et al. 1999; Ikeda et al. 1996; Mangiarini et al. 1996) carrying expanded repeat transgenes but intact endogenous alleles show phenotypes resembling the respective diseases. Together these data suggest a gain-of-function mechanism of pathogenesis. Similar disease severity in affected siblings who are either heterozygous or homozygous for the mutant HD chromosome also suggested the neomorphic nature of the disease mutation (Wexler et al. 1987). The generation of *Drosophila* models of this group of diseases is a logical extension of these observations, since the specific human gene causing the disorder can be introduced into the fly genome as a transgene and can be reproducibly targeted to different cell types using the GAL4/UAS system for ectopic gene expression (Brand and Perrimon 1993).

HD is characterized by chorea, cognitive impairment, changes in affect, and diffuse degeneration of the neostriatum, with onset usually occurring during adulthood (Gusella and MacDonald 1995). The human HD gene encodes huntingtin (Htt), a novel 350 kd cytoplasmic protein expressed throughout the brain (Trottier et al. 1995). Dynamic mutations within exon 1 can increase the size of the normal allele comprised of a polyglutamine tract of 37 or fewer CAG repeats to a disease-causing allele with greater than 150 repeats (Gusella and MacDonald

1995). Jackson et al. (1998) expressed N-terminal fragments of human huntingtin containing tracts of 2, 75, and 120 glutamine residues in photoreceptor neurons in the fly eye using the gmr enhancer. Whereas wild-type flies and flies carrying two glutamine (Q2) repeats showed normal eye morphology in the adult, loss of rhabdomeres was seen in Q75 flies by day 10 and degeneration was evident by two days if 120 glutamines (Q120) were present in the protein. This is similar to human HD, where the length of the polyglutamine repeat determines the age of onset, so that patients with juvenile onset generally encode proteins with repeat lengths of greater than 70 glutamine residues (Gusella and MacDonald 1995). The proteins with 2, 75, or 120 CAG repeats were all found in the cytoplasm in developing neurons in the larval eye primordium followed by progressive nuclear accumulation of the Q75 or Q120 proteins. At 10 days of age, Q75 was found in the nuclei of scattered photoreceptor neurons and Q120 was found predominantly within the nucleus in multiple small aggregates. The correlation between nuclear localization of the glutamine repeat-containing proteins and neuronal degeneration found in some striatal and cortical neurons in HD patients (Cummings and Zoghbi 2000) and in transgenic mice (Scherzinger et al. 1997) is also confirmed in the *Drosophila* HD model.

A toxic novel function for the mutant proteins was also suggested by the appearance of ubiquitin as well as huntingtin in neuronal inclusions (NI) from mice transgenic for exon 1 of huntingtin (Davies et al. 1997). The presence of nuclear or cytoplasmic aggregates was later validated by examination of post-mortem tissue from patients with eight polyglutamine diseases (Cummings and Zoghbi 2000) and in other transgenic mouse models (Skinner et al. 1997). Importantly, the presence of ubiquitin, as well as chaperones and components of the proteasome (Chai et al. 1999a,b; Cummings et al. 1998), suggested that polyglutamine expansion leads to abnormal protein conformation, and that cellular stress pathways are activated to refold or degrade the mutant proteins or to modify protein aggregation.

A role for nuclear localization in neurodegeneration, independent of the formation of neuronal aggregates, was demonstrated in mice with a SCA1 allele bearing a mutated nuclear localization signal (SCA1 82Q K772T). High levels of ataxin-1 expressed in the cytoplasm rather than in the nucleus of Purkinje cells did not result in either ataxia or neuropathology in these mice (Klement et al. 1998). The discovery of altered expression of neuronal genes before detectable pathology in SCA1 transgenic mice (Lin et al. 2000) and of gene transcription mediated by CBP in HD transgenic mice (Nucifora et al. 2001) suggests that the aberrant localization may facilitate abnormal interactions with the transcriptional machinery. Expanded polyglutamine proteins have been found to cause the redistribution of key transcription factors such as nuclear receptor co-repressor (N-CoR) and CREB-binding protein (CBP) in mouse models, cell culture models, and human tissues; the sequestration of such factors might have detrimental effects on gene regulation (Nucifora et al. 2001; Shimohata et al. 2000; Boutell et al. 1999; Steffan et al. 2000; McCampbell et al. 2000; Yvert et al. 2001). Notably, several enhancers of SCA1 pathogenesis in *Drosophila* were transcriptional cofactors, including Sin3a, Rpd3 and dCtBP (Fernandez-Funez et al. 2000).

Studies in *Drosophila* models of SCA1 (Fernandez-Funez et al. 2000) and SCA3/MJD (Warrick et al. 1998) have revealed that differences in expression levels of a mutant protein as well as the length of the protein containing an expanded polyglutamine tract can result in major differences in disease severity. Normal and defective variants of *MJD1*, a ubiquitously expressed gene causing SCA3/MJD, were introduced as transgenes in *Drosophila* to analyze the mechanisms of neuronal pathology and cellular specificity observed in this disease (Warrick et al. 1998). MJD is characterized by bulging eyes, facial and lingual fasciculation, rigidity, progressive ataxia, and degeneration of several regions of the brain. These include the basal ganglia, brain stem, spinal cord, and dentate neurons of the cerebellum, as well as mild loss of Purkinje cells (Durr et al. 1996). The polyglutamine repeat in *MJD1* is in the C-terminal of the protein and varies from between 12 and 40 glutamine residues in normal alleles to a repeat length between 55 and 84 glutamines for affected chromosomes (Kawaguchi et al. 1994). A C-terminal fragment containing 78 glutamine repeats was directed to all developing cells of the nervous system and resulted in embryonic lethality when highly expressed but caused early death in adulthood if weakly or moderately expressed. Severity was also dependent on the strength of the transgene in the larval eye disc, where nuclear inclusions formed more rapidly and were more prominent in the transgenic lines with enhanced expression of the 78Q protein. A control transgene containing 27 glutamine repeats had no phenotypic effect even if highly expressed when assayed under the same conditions.

The association between level of transgene expression and phenotypic severity was again seen in the *Drosophila* SCA1 model after introduction of full-length ataxin-1 transgenes differing in the length of the polyglutamine tract (Fernandez-Funez et al. 2000). The clinical features of SCA1 include progressive cerebellar ataxia, dysarthria, bulbar dysfunction, and severe loss of Purkinje cells. Normal alleles encoding ataxin-1 contain 6-44 repeats, whereas disease alleles bear an uninterrupted stretch of CAG trinucleotides ranging from 39–82 residues (Zoghbi and Orr 1995). Unexpectedly, high expression of both a wild-type human isoform including 30 glutamine repeats and ataxin-1 82Q caused retinal degeneration and the formation of nuclear inclusions, although the 82Q transgenic lines showed a stronger phenotype.

A *Drosophila* model has also been used to address the intrinsic cytotoxicity of long chains of polyglutamines that may be independent of their disease gene context. Ordway et al. had previously engineered 146 CAG repeats into the mouse *Hprt* gene to determine whether the expanded glutamine tract had to be embedded within one of the known repeat disorder genes to exert a pathological effect. The mutant mice developed late-onset neurological dysfunction, including tremors, seizures, and motor dysfunction, and died prematurely. In addition, there was evidence of neuronal intranuclear inclusions containing *Hprt*, a protein normally found in the cytoplasm, but there was no selective neuronal loss. These observations indicated that the polyglutamine tract itself was neurotoxic (Ordway et al. 1997). For the fly experiments, Marsh et al. constructed transgenes containing either 22 or 108 glutamine residues flanked by 6 or 4 amino acids on the N- and C- terminal sides, respectively. Expression of the Q22 peptide had no effect on viability, whereas the Q108 protein caused nearly complete lethality when under the control of a variety of cell-type-specific promoters. There was also evidence of neurodegeneration that was cell-type dependent, so that photo-receptor neurons but not mechanosensory hair cells were affected in flies harbor-ing the Q108 transgene. Toxicity was reduced in all cases by the addition of myc or flag epitope tags 26 amino acids in length to the peptides containing 108 CAG repeats. These results also suggest that the polyglutamine tract alone was delete-rious but that the phenotypic consequences can be modified by protein context, possibly due to changes in protein solubility or aggregation (Marsh et al. 2000).

## Parkinson's Disease

Parkinson's disease (PD) is a dominantly inherited neurodegenerative syndrome associated with loss of dopaminergic neurons in the substantia nigra pars com-pacta and the locus coeruleus. PD is characterized by muscle rigidity, resting tremor, slowed movement and poor balance, and in some cases by compromised cognitive function and memory. The hallmark of PD at the cellular level is the formation of intracytoplasmic protein aggregates either in the cytoplasm of neuronal cell bodies or neuronal processes, known as Lewy bodies or Lewy neurites, respectively (Sethi 2002). The aggregates contain synphilin-1, ubiquitin, parkin (an E3 ubiquitin ligase), as well as α-synuclein (Okochi et al. 2000) as major constituents. Although the function of α-synuclein has not been clearly defined, the neurochemical, electrophysiological and behavioral deficits evident in mice carrying a null mutation (Abeliovich et al. 2000; Clayton and George 1998) and its localization in presynaptic neurons (deSilva et al. 2000) suggest a possible role for α-synuclein in synaptic plasticity. Lewy bodies and Lewy neurites containing α-synuclein are also found in a group of neurodegenerative disorders known as synucleinopathies (Spillantini et al. 1997, 1998a; Baba et al 1998), including dementia with Lewy bodies (DLB), Lewy body variant of Alzhe-imer's disease (LBVAD) and Multiple System Atrophy (MSA). Lewy bodies and β-amyloid deposits occur throughout the brain in both DLB and LVD, whereas the defining neuropathological feature of MSA is α-synuclein glial cytoplasmic inclusions (Duda et al. 2000).

Although most cases of PD are sporadic, rare familial cases of PD have been reported. Polymeropoulos et al. (1997)described a mutation in α-synuclein seg-regating in an Italian-Greek family in which alanine was replaced by threonine at amino acid position 53 of the protein (A53T). Similarly, a mutation involving the substitution of alanine for phenylalanine at position 30 (A30P) was later identi-fied in a German family (Kruger et al. 1998), suggesting a direct link between α-synuclein and disease pathogenesis. In vitro experiments showing that α-synu-clein can self-aggregate and form fibers (Conway et al. 1998; Narhi et al. 1999) may further indicate that the mutations in α-synuclein reported in these families are likely to be implicated in the formation of the pathognomonic Lewy bodies and Lewy neurites.

A model of PD in which the human disease is largely recapitulated was devel-oped in *Drosophila* by overexpression of either wild-type human α-synuclein or the two mutant forms of the gene identified in the familial PD cases (Feany and

Bender 2000). Alpha-synuclein expression was directed to a variety of cell types using the GAL4-UAS system. Flies expressing wild-type or mutant α-synuclein cDNAs were viable and showed normal development of both neuronal and non-neuronal tissues at eclosion. However, an age-dependent loss of dorsomedial dopaminergic neurons was found in the absence of widespread degenerative changes in the brain in 30- to 60-day-old flies expressing α-synuclein in a pan-neural pattern. Serotonergic cells examined as an index of the specificity of the observed neurodegeneration were unaffected. The dorsomedial cluster of neurons was also depleted at 30 days in flies expressing wild-type, A30P or A53T α-synuclein under the control of the promoter for the DOPA decarboxylase gene, another marker of dopaminergic neurons. Substantial numbers of dopaminergic neurons still remained in both types of aged transgenic flies, suggesting a possible selective vulnerability of subsets of neurons to α-synuclein toxicity, as is seen in PD patients. Formation of cytoplasmic α-synuclein fibrillar deposits that resembled human cortical Lewy bodies and Lewy neurites was apparent at 20–30 days of age, when mutant α-synuclein expression was pan-neural. A progressive decline in climbing ability that was slightly more rapid than the normal age-related changes seen in control flies was a further reflection of nervous system dysfunction in the α-synuclein expressing flies. The loss of motor control was most severe in flies bearing the A30P mutation, but variations in the level of expression of the various transgenes may explain the apparent differences.

It appears thus far that the PD *Drosophila* model most closely resembles the human disease. Transgenic mice overexpressing α-synuclein show dopaminergic neuron dysfuncton and loss and the development of intracytoplasmic neuronal α-synuclein inclusions with (Giasson et al. 2002) or without fibrillar deposits (Masliah et al. 2000; van der Putten et al. 2000; Kahle et al. 2000; Lee et al. 2002). None of the published PD mouse models exhibit pathological lesions in tyrosine hydrolase-expressing neurons of the substantia nigra. One likely explanation of this discrepancy is that expression of α-synuclein in the current mouse models is not adequate to permit sufficient accumulation in the relevant neurons during the life span of the mouse. Another possibility is that endogenous synucleins in mice but not in flies might retard neurodegeneration. Mice express three neuronal protein isoforms of synuclein (α-, β-, γ-). β-synuclein inhibits α-synuclein aggregation in cell lines and suppresses α-synuclein pathology in transgenic mice (Hashimoto et al. 2001). The presence of endogenous murine α-synuclein could also inhibit the aggregation or mitigate the pathological effects of the A53T mutation carried on a transgene. This hypothesis could be tested in mice by crossing the α-synuclein null mice with α-synuclein transgenic mice and by using the existing *Drosophila* PD model to evaluate the effect of overexpressing β- or γ-synuclein.

Autosomal recessive juvenile parkinsonism (AR-JP), characterized by early onset and absence of Lewy bodies, is associated with recessive mutations in the parkin gene (Kitada et al. 1998). Parkin encodes an E3 ubiquitin ligase (Shimura et al. 2000) that has been reported to promote the ubiquitination of the synaptic-vesicle-associated protein CDCrel-1 (Zhang et al. 2000), synphilin-1, an α-synuclein interacting protein (Chung et al. 2001, and α-Sp22, an O-glycosylated isoform of α-synuclein (Shimura et al. 2001). Parkin expression levels increase after cellular stress, so that one of its functions could be to remove misfolded proteins in the endoplasmic reticulum where it also suppresses cell death associated with stress (Imai et al. 2001). These observations have led to the formulation of the widely accepted hypothesis that lack of parkin activity would lead to the accumulation of its substrates and eventual cellular toxicity (Giasson and Lee 2001). The introduction of transgenes overexpressing parkin or of mutations abrogating parkin expression in flies would offer a means to test this hypothesis. Altering parkin activity and investigating the phenotypic consequences in the current *Drosophila* PD models would be another alternative set of experiments that would provide insight into PD pathogenesis.

## Tauopathies

The tauopathies are a family of neurodegenerative diseases that have in common the formation of filamentous intracellular deposits of the tau protein. In Alzheimer's disease (AD), the most well known of the tauopathies and the most common neurodegenerative disease, extracellular β-amyloid plaques are found together with intracytoplasmic neurofibrillary tangles (NFT) containing hyperphosphorylated tau. Other members of this family of diseases include progressive supranuclear palsy (PSP), corticobasal degeneration (CBD), Pick's disease (PiD), and frontotemporal dementia with parkisonism (FTDP-17). Mutations in the *MAPT* gene that encodes tau have been identified as the cause of a small proportion of cases of autosomal dominant familial FTDP-17 (Hutton et al. 1998; Spillantini et al. 1998b; Poorkaj et al. 1998), showing that abnormal tau can directly cause neurodegeneration.

Tau accumulates primarily in the axons of healthy neurons where it plays an important role in microtubule stability by binding and connecting microtubules and polymerizing actin. Alternative splicing of the *MAPT* transcript results in the production of six protein isoforms in adult brain (Goedert et al. 1989). Mutations in the splice donor site of exon 10 in FTDP-17 patients are associated with predominant parkinsonism and result in a relative increase of a splice form containing four microtubule binding repeat domains and a decrease in an isoform with three repeat domains. Mutations in exons 9, 10, 12, and 13 reduce microtubule binding and lead to predominant dementia (Grover et al. 1999; Reed et al. 2001). Mutations in *MAPT* have not been found in AD, despite the presence of the hallmark NFT, but mutations in regulatory elements or defective protein interactions cannot be excluded.

In disease conditions, tau is hyperphosphorylated and shows a decreased ability to bind to microtubules. The increased phosphorylation may therefore explain the mislocalization of tau to the somatodendritic cellular compartment, where it aggregates in filamentous deposits (Drechsel et al. 1992; Bramblett et al. 1993; Yoshida and Ihara 1993; Biernat et al. 1993). Tau might be phosphorylated by different kinases in different populations of neurons (Gotz and Nitsch 2001), but cyclin-dependent kinase (cdk-5) and glycogen synthase kinase-3β (GSK-3β) have been shown to interact with tau in vivo. Mice that overexpress human p25, an activator of cdk-5, display tau hyperphosphorylation and cytoskeletal disruption reminiscent of AD but without filamentous deposits (Ahlijanian et al. 2000). Similarly, overexpression of GSK-3β in transgenic mice causes tau hyperphosphorylation and somatodendritic localization, as well as neuronal apoptosis (Lucas et al. 2001). These two studies suggest that tau hyperphosphorylation alone is sufficient to cause neurodegeneration.

Tau transgenic mice overexpressing the FTDP-17 mutations P301L, V337M, or G272V show motor and behavioral deficits as well as age-dependent development of NFT, but not the selective neuronal loss characteristic of AD (Lewis et al. 2000; Gotz et al. 2001a,b; Tanemura et al. 2001). Mice expressing wild-type tau from a number of different promoters show somatodendritic accumulation of tau suggestive of pre-tangle pathology but do not form NFT, supporting the idea that the FTDP-17 mutations promote tau aggregation (Gotz et al. 1995; Brion et al. 1999; Ishihara et al. 1999; Spittaels et al. 1999; Probst et al. 2000).

The recent development of *Drosophila* models of tauopathies offers the possibility of using genetic screens to unravel the mechanisms of pathogenesis. Wittmann et al. (2001) first reported that transgenic flies expressing wild-type or FTDP-17 mutant tau (R406W or V337M) show a shortened life span when the tau protein was overexpressed. The nervous system of the transgenic flies showed no developmental or anatomical abnormalities at eclosion but showed increasing vacuolization and degeneration of cortical neurons by day 20. Both of these phenotypes are more severe with mutant tau, but differences in expression levels from the different transgenes cannot be excluded to explain the differences. Monoclonal antibodies recognizing either hyperphosphorylated tau or tau with abnormal conformations present in brain tissue from patients with degenerative disease were used to confirm the age-dependent accumulation of defective protein in the brain of transgenic flies. In addition, abnormal tau was concentrated in areas of neurodegeneration. When tau expression was directed to cholinergic neurons, progressive cell loss was observed that was exacerbated in the R406W transgenic fly lines. These phenotypes occurred, however, in the absence of detectable NFT. Because NFT occur in areas of the human brain that undergo neurodegeneration, it has been assumed that fibrillar aggregates are the cause of the tauopathies, either by acting as physical barriers to normal cell function or by exerting toxic effects. On the basis of their results, Wittmann et al. suggest that neurotoxicity may not depend on the formation of large aggregates but on protein alterations that occur before the appearance of NFT. These modifications would be presumed to include the conformational changes and hyperphosphorylation observed in *Drosophila*. Although the authors do not dismiss the possibility of a species difference in the pathways of neurodegeneration, the extensive neurodegeneration observed in FTDP-17 patients who show few NFT (Bird et al. 1999) is consistent with a separation between tangle formation and tau toxicity.

A second *Drosophila* model of neurodegeneration induced by wild-type tau was recently reported (Jackson et al. 2002). Jackson et al. tested for interactions between tau and the Wnt signal transduction pathway by overexpressing *shaggy*, the *Drosophila* GSK-3β homolog. If Wnt signaling occurs, GSK-3β is inhibited, leading to decreased phosphorylation of Armadillo, the *Drosophila* homolog of beta-catenin. Armadillo is subsequently translocated to the nucleus where it acts as a transcriptional coactivator with dTCF/ pangolin (Moon et al. 2002). Introduction of a wild-type tau transgene (htau4R) in the retina resulted in degeneration without the appearance of NFT, as described by Wittmann and colleagues (2001). Coexpression of tau and of *shaggy* in a pan-neural pattern caused an exacerbation of the rough eye phenotype and disordered ommatidia induced by tau, whereas tau expression in a background heterozygous for a *shaggy* hypomorphic allele suppressed neurodegeneration. In addition, neurofibrillary aggregates containing hyperphosphorylated tau were detected by immunohistochemistry in the photoreceptor neurons of flies expressing tau in the presence of *shaggy*. By generating mutations in both *armadillo* and dTCF, components of the Wnt pathway that act downstream of GSK-3β, and analyzing the resulting phenotypes, Jackson et al. concluded that GSK-3β did not exert its effects by antagonizing the Wnt pathway. For example, loss of function mutations in both *armadillo* and TCF that would be expected to reduce Wnt signaling suppressed rather than enhanced cellular disorganization in the eye. These observations are most consistent with the interpretation that GSK-3βdirectly phosphorylates tau and that tau hyperphosphorylation can cause NFT-like pathology in vivo.

To date, there are no reported β-amyloid fly models for AD. The development of these models will likely further contribute to the understanding of the pathogenesis of AD.

## Drosophila Models and Development of Therapy for Neurodegenerative Disease

The development of effective strategies for the treatment of neurodegenerative diseases has so far not been attainable. Recent advances in understanding the mechanisms involved in the pathogenesis of neurodegenerative disease may lead to identification of specific molecular targets amenable to drug treatment. Remarkably, recovery of neuronal and motor function and the reversal of aggregation of mutant protein were observed when expression of a tetracycline-inducible mutant huntingtin fragment was discontinued in symptomatic mice (Yamamoto et al. 2000). These experiments suggest that HD or other neurodegenerative diseases may be reversible.

Studies in both flies and mice have shown that overexpression of chaperone or heat shock proteins, which help refold proteins or target them for degradation, increases resistance to polyglutamine-induced toxicity. Overexpression of the human gene encoding Hsp70 restores retinal structure and improves the survival of SCA3 transgenic flies overexpressing ataxin-3 containing 78 glutamine residues (Warrick et al. 1999). The suppression of protein toxicity was not accompanied by a change in the formation of aggregates, except that a fraction of the exogenous HSP70 was recruited to the NI, indicating that Hsp70 may interact directly with the disease protein. When the same transgenic flies inherit an endogenous Hsp70 (Hsc4) dominant-negative mutant allele, more severe degeneration is observed.

Genetic screens in the SCA1 fly model revealed that loss-of-function mutations in protein-folding pathway components enhanced neurodegeneration and, conversely, overexpression of a DNAJ1 chaperone mitigates neuronal loss (Fernandez-Funez et al. 2000). Mutations inactivating ubiquitin, or a ubiquitin carboxy terminal hydrolase accelerate neurodegeneration, confirming the importance of protein clearance in pathogenesis. A genetic screen to detect genes that dominantly modify polyglutamine toxicity was also undertaken in transgenic flies carrying synthetic polyglutamine tracts with repeat lengths of either 20 or 127 CAG residues (Kazemi-Esfarjani and Benzer 2000). Flies bearing transgenes with 127 glutamine repeats had marked external eye abnormalities, so that the screen entailed isolating genes from P-element chromosomal insertion sites that were associated with suppression or enhancement of degeneration. Two of these suppressors, *Drosophila* homologues of the molecular chaperone Hsp40 and the human tetratricopeptide repeat protein 2, were confirmed by testing the corresponding cDNAs in transgene constructs injected into fly embryos.

As was seen in the *Drosophila* polyglutamine disease models, human Hsp70 can modify α-synuclein toxicity in the fly. Overexpression of HSP70 inhibits dopaminergic neuronal loss caused by α-synuclein under the control of the DOPA decarboxylase promoter, and reduction of Hsc4 activity intensifies α-synuclein toxicity. The involvement of HSP70 in PD is also suggested by the observation that 1-5% of Lewy bodies show HSP70- or HSP40-positive immunostaining in human postmortem tissue (Auluck et al. 2002).

Chaperone overexpression could be protecting neurons at multiple levels. The addition of exogenous chaperones may improve protein solubility and clearance, as reflected in the amelioration of neurodegeneration. This type of biochemical change in polyglutamine protein was observed by Chan et al. in SCA3 transgenic flies coexpressing ataxin-3-containing 78 glutamine repeats and either human Hsp70 or a fly ortholog of a human Hsp40 chaperone (Chan et al. 2000). When neurodegeneration was suppressed by chaperones, the solubility properties of the mutant polyglutamine protein appeared to be altered, as reflected in the ability to be extracted by either high salt or SDS. Molecular chaperones might also confer protection by delaying neurodegeneration through inhibition of signal transduction pathways leading to cell death. This could take place by a number of mechanisms, including prevention of activation of stress kinases (Gabai et al. 1997), by blocking procaspase processing or caspase activation (Mosser et al. 2000; Zhou et al. 2001), or by preventing the interaction between misfolded proteins and other cellular factors. It is also possible that polyglutamine proteins or α-synuclein could mediate neurotoxicity by interfering with endogenous chaperone activity, so that the expression of exogenous chaperones could overcome this effect(Warrick et al. 1999; Auluck et al 2002).

In most of these studies performed in the fly, toxicity was suppressed without any noticeable effect on the formation of protein aggregates, but see Fernandez-Funez et al. (2000) for an exception. Chaperones may have an effect on protein conformation or levels in addition to an effect on solubility that is not easy to detect when the disease-causing genes are expressed at very high levels in the transgenic models. While the role of the protein aggregates in pathogenesis is not entirely clear, some of the data argue that aggregates are not necessary to initiate pathogenesis. In SCA1 mice, loss of function of *Ube3a*, an E3 ubiquitin ligase, intensifies neurodegeneration, even though it decreases the number of nuclear aggregates (Cummings et al. 1999). In a SCA1 mouse model in which 154 glutamine repeats are targeted to the endogenous *ataxin-1* locus, the most vulnerable neurons were the last to form visible aggregates, whereas those that were spared formed aggregates earlier (Watase et al. 2002). Based on these two studies, it appears that aggregate formation may sequester the protein and curtail its toxic effects. Because many of the neurodegenerative diseases have the accumulation of insoluble toxic structures as a common feature, modulation of chaperone activity may be of general benefit for this type of disorder.

The efficacy of two other therapeutic approaches for neurodegenerative disorders has also been tested using *Drosophila* models. Since mutant huntingtin interferes with the acetyltransferase activity of the enzymes CBP (CREB-binding protein), P/CAF and p300, and reduces the level of acetylated histones H3 and H4 in vitro, polyglutamine pathogenesis might involve a decrease in histone acetylation. In addition, overexpression of CBP reduces polyglutamine-mediated death of cultured cells, presumably by reversing loss of histone acetylation when CBP is sequestered in nuclear and cytoplasmic aggregates (Nucifora et al. 2001; Steffan et al. 2000; McCampbell et al. 2000). To examine the effect of raising global levels of acetylation, transgenic flies expressing either the polyglutamine-containing domain of Huntingtin with 93 glutamine repeats (Httex1p Q93) or a peptide comprised of an extended tract of 48 CAG residues (Q48) were fed with sodium butyrate or suberoylanilide hydroxamic acid (SAHA; Steffan et al. 2001), two compounds that are thought to work by inhibiting histone deacetylase (HDAC) and perhaps other proteins. Expression of Httex1p Q93 or Q48 in photoreceptor neurons led to a progressive loss of rhabdomeres that was ameliorated by treatment with both HDAC inhibitors even after the onset of symptoms. To further test the hypothesis that interference with HDAC activity can ameliorate neuronal degeneration, levels of HDAC were reduced genetically by a heterozygous partial loss of function mutation of *Sin3A*, a component of HDAC complexes. Reduced activity of *Sin3A* was shown to increase the viability of Httex1p Q93 flies and decrease progressive neurodegeneration (Steffan et al. 2001). In contrast, reducing the activity of HDACs in a fly model of SCA1 worsened neurodegeneration. These distinct findings might reflect differences in the mechanisms of pathogenesis among polyglutamine diseases. In this context, it will be interesting to determine the effect of altering the levels of the CBP acetyltransferase in the SCA1 and HD (Httex1p Q93) fly models.

The prevention of aberrant protein interactions is another possible strategy to develop a therapy for neurodegenerative diseases. Kazantsev et al. (2002) showed that a synthetic, bivalent Htt-binding peptide inhibited aggregation in cell culture by co-localizing with and binding directly to a truncated Htt sequence containing 103 CAG repeats. Coexpression of the suppressor and expanded polyglutamine polypeptides containing either 48 or 108 CAG repeats in transgenic flies (Marsh et al. 2000) increased survival and reduced neurodegeneration observed in photoreceptors in the eye. Reduction of pathology was correlated with decreased aggregate formation in vivo in neuroblasts and mature neurons in the third-instar larval brain and in the eye imaginal disc. These experiments suggest that small molecules designed to interfere with the aggregation of mutant protein in large inclusions may eventually be useful in the treatment of HD and other neurodegenerative disorders.

## Conclusions and future directions

*Drosophila* models of neurodegenerative disease have provided essential insights into pathological mechanisms and genetic pathways that appear to be at least partly conserved between vertebrates and invertebrates. Of particular value was the observation in SCA1 and PD transgenic flies that high expression of wild-type ataxin-1 or α-synuclein could result in the appearance of pathological features of the human disorders. Fernandez-Funez et al. present a compelling argument to explain this observation in the SCA1 model, speculating that a certain percentage of even wild-type ataxin-1 has a tendency to misfold and that expansion of the polyglutamine tract will serve to maximize this likelihood. The cell's proteolytic machinery is able under normal physiological conditions to degrade the abnormal ataxin-1 protein but loses this capability when the amount of the defective protein exceeds a critical level (Fernandez-Funez et al. 2000). Other novel findings include evidence for dissociation between tangle formation and tau neurotoxicity in a fly model of tauopathy (Wittmann et al. 2001) and the establishment of a role for GSK-3βin tau phosphorylation that does not involve the canonical Wnt signal transduction pathway (Jackson et al. 2002).

The availability of fly models that closely resemble the human disorders is an invaluable resource for performing genetic screens to identify suppressors and enhancers of neurodegeneration that may eventually serve as protein or transcriptional targets for the design of new drugs. This system has already been used productively for both the SCA1 and HD *Drosophila* models, leading to the identification of endogenous molecular chaperones (Kazemi-Esfarjani and Benzer 2000) and new mechanisms of polyglutamine pathogenesis with the discovery of modifiers that are involved in RNA processing, transcriptional regulation and cellular detoxification (Fernandez-Funez et al. 2000). Similarly, the *Drosophila* models promise to have great utility for both hypothesis testing and screening of new therapeutic agents. A recent paper describing treatment of transgenic flies expressing mutant Huntingtin protein with HDAC inhibitors to offset transcriptional repression is probably only the first of many studies that will exemplify this approach (Steffan et al. 2001).

## Acknowledgments

I thank Jan Bressler for outstanding editorial assistance. Work in my laboratory is supported by the National Institutes of Health.

## References

Abeliovich A, Schmitz Y, Farinas I, Choi-Lundberg D, Ho WH, Castillo PE, Shinsky N, Verdugo JM, Armanini M, Ryan A, Hynes M, Phillips H, Sulzer D, Rosenthal A (2000) Mice lacking alpha-synuclein display functional deficits in the nigrostriatal dopamine system. Neuron 25: 239-252

Adams MD, Celniker SE, Holt RA, Evans CA, Gocayne JD, Amanatides PG, Scherer SE, Li PW, Hoskins RA, Galle RF, George RA, Lewis SE, Richards S, Ashburner M, Henderson SN, Sutton GG, Wortman JR, Yandell MD, Zhang Q, Chen LX, Brandon RC, Rogers YH, Blazej RG, Champe M, Pfeiffer BD, Wan KH, Doyle C, Baxter EG, Helt G, Nelson CR, Gabor GL, Abril JF, Agbayani A, An HJ, Andrews-Pfannkoch C, Baldwin D, Ballew RM, Basu A, Baxendale J, Bayraktaroglu L, Beasley EM, Beeson KY, Benos PV, Berman BP, Bhandari D, Bolshakov S, Borkova D, Botchan MR, Bouck J, Brokstein P, Brottier P, Burtis KC, Busam DA, Butler H, Cadieu E, Center A, Chandra I, Cherry JM, Cawley S, Dahlke C, Davenport LB, Davies P, de Pablos B, Delcher A, Deng Z, Mays AD, Dew I, Dietz SM, Dodson K, Doup LE, Downes M, Dugan-Rocha S, Dunkov BC, Dunn P, Durbin KJ, Evangelista CC, Ferraz C, Ferriera S, Fleischmann W, Fosler C, Gabrielian AE, Garg NS, Gelbart WM, Glasser K, Glodek A, Gong F, Gorrell JH, Gu Z, Guan P, Harris M, Harris NL, Harvey D, Heiman TJ, Hernandez JR, Houck J, Hostin D, Houston KA, Howland TJ, Wei MH, Ibegwam C, Jalali M, Kalush F, Karpen GH, Ke Z, Kennison JA, Ketchum KA, Kimmel BE, Kodira CD, Kraft C, Kravitz S, Kulp D, Lai Z, Lasko P, Lei Y, Levitsky AA, Li J, Li Z, Liang Y, Lin X, Liu X, Mattei B, McIntosh TC, McLeod MP, McPherson D, Merkulov G, Milshina NV, Mobarry C, Morris J, Moshrefi A, Mount SM, Moy M, Murphy B, Murphy L, Muzny DM, Nelson DL, Nelson DR, Nelson KA, Nixon K, Nusskern DR, Pacleb JM, Palazzolo M, Pittman GS, Pan S, Pollard J, Puri V, Reese MG, Reinert K, Remington K, Saunders RD, Scheeler F, Shen H, Shue BC, Siden-Kiamos I, Simpson M, Skupski MP, Smith T, Spier E, Spradling AC, Stapleton M, Strong R, Sun E, Svirskas R, Tector C, Turner R, Venter E, Wang AH, Wang X, Wang ZY, Wassarman DA, Weinstock GM, Weissenbach J, Williams SM, WoodageT, Worley KC, Wu D, Yang S, Yao QA, Ye J, Yeh RF, Zaveri JS, Zhan M, Zhang G, Zhao Q, Zheng L, Zheng XH, Zhong FN, Zhong W, Zhou X, Zhu S, Zhu X, Smith HO, Gibbs RA, Myers EW, Rubin GM, Venter JC (2000) The genome sequence of Drosophila melanogaster. *Science 287: 2185–2195*

Ahlijanian MK, Barrezueta NX, Williams RD, Jakowski A, Kowsz KP, McCarthy S, Coskran T, Carlo A, Seymour PA, Burkhardt JE, Nelson RB, McNeish JD (2000) Hyperphosphorylated tau and neurofilament and cytoskeletal disruptions in mice overexpressing human p25, an activator of cdk5. Proc Natl Acad Sci USA 97: 2910–2915

Auluck PK, Chan HY, Trojanowski JQ, Lee VM, Bonini NM (2002) Chaperone suppression of alpha-synuclein toxicity in a Drosophila model for Parkinson's disease. Science295: 865–868

Baba M, Nakajo S, Tu PH, Tomita T, Nakaya K, Lee VM, Trojanowski JQ, Iwatsubo T (1998) Aggregation of alpha-synuclein in Lewy bodies of sporadic Parkinson's disease and dementia with Lewy bodies. Am J Pathol 152: 879–884

Biernat J, Gustke N, Drewes G, Mandelkow EM, Mandelkow E (1993) Phosphorylation of Ser262 strongly reduces binding of tau to microtubules: distinction between PHF-like immunoreactivity and microtubule binding. Neuron11: 1531–1563

Bird TD, Nochlin D, Poorkaj P, Cherrier M, Kaye J, Payami H, Peskind E, Lampe TH, Nemens E, Boyer PJ, Schellenberg GD (1999) A clinical pathological comparison of three families with frontotemporal dementia and identical mutations in the tau gene (P301L). Brain 122: 741–756

Boutell JM, Thomas P, Neal JW, Weston VJ, Duce J, Harper PS, Jones AL (1999) Aberrant interactions of transcriptional repressor proteins with the Huntington's disease gene product, huntingtin. Human Mol Genet 8: 1647–1655

Bramblett GT, Goedert M, Jakes R, Merrick SE, Trojanowski JQ, Lee VM (1993) Abnormal tau phosphorylation at Ser396 in Alzheimer's disease recapitulates development and contributes to reduced microtubule binding. Neuron 10: 1089–1099

Brand AH, Perrimon N (1993) Targeted gene expression as a means of altering cell fates and generating dominant phenotypes. Development 118: 401–415

Brion JP, Tremp G,Octave JN (1999) Transgenic expression of the shortest human tau affects its compartmentalization and its phosphorylation as in the pretangle stage of Alzheimer's disease. Am J Pathol 154: 255–270

98    Juan Botas

Burright EN, Clark HB, Servadio A, Matilla T, Feddersen RM, Yunis WS, Duvick LA, Zoghbi HY, Orr HT (1995) SCA1 transgenic mice: a model for neurodegeneration caused by an expanded CAG trinucleotide repeat. Cell 82: 937–948

Chai Y, Koppenhafer SL, Bonini NM, Paulson HL (1999a) Evidence for proteasome involvement in polyglutamine disease: localization to nuclear inclusions in SCA3/MJD and suppression of polyglutamine aggregation in vitro. Human Mol Genet 8: 673–682

Chai Y, Koppenhafer SL, Shoesmith SJ, Perez MK, Paulson HL (1999b) Analysis of the role of heat shock protein (Hsp) molecular chaperones in polyglutamine disease. J Neurosci 19: 10338–10347

Chan HY, Warrick JM, Gray-Board GL, Paulson HL, Bonini NM (2000) Mechanisms of chaperone suppression of polyglutamine disease: selectivity, synergy and modulation of protein solubility in Drosophila. *Human Mol Genet 9: 2811–2820*

Chung KK, Zhang Y, Lim KL, Tanaka Y, Huang H, Gao J, Ross CA, Dawson VL, Dawson TM (2001) Parkin ubiquitinates the alpha-synuclein-interacting protein, synphilin- 1: implications for Lewy-body formation in Parkinson disease. Nature Med 7: 1144–1150

Clayton DF,George JM (1998) The synucleins: a family of proteins involved in synaptic function, plasticity, neurodegeneration and disease. Trends Neurosci 21: 249–254

Conway KA, Harper JD, Lansbury PT (1998) Accelerated in vitro fibril formation by a mutant alpha-synuclein linked to early-onset Parkinson disease. Nature Med 4: 1318–1320

Cummings CJ, Mancini MA, Antalffy B, DeFranco DB, Orr HT, Zoghbi HY (1998) Chaperone suppression of aggregation and altered subcellular proteasome localization imply protein misfolding in SCA1. Nature Genet 19: 148–154

Cummings CJ, Reinstein E, Sun Y, Antalffy B, Jiang Y, Ciechanover A, Orr HT, Beaudet AL, Zoghbi HY (1999) Mutation of the E6-AP ubiquitin ligase reduces nuclear inclusion frequency while accelerating polyglutamine-induced pathology in SCA1 mice. Neuron 24: 879–892

Cummings CJ, Zoghbi HY (2000) Trinucleotide repeats: mechanisms and pathophysiology. Annu Rev Genomics Human Genet 1:281–328

Davies SW, Turmaine M, Cozens BA, DiFiglia M, Sharp AH, Ross CA, Scherzinger E, Wanker EE, Mangiarini L, Bates GP (1997) Formation of neuronal intranuclear inclusions underlies the neurological dysfunction in mice transgenic for the HD mutation. Cell 90: 537–548

de Silva HR, Khan NL, Wood NW (2000) The genetics of Parkinson's disease. Curr Opin Genet Dev 10: 292–298

Drechsel DN, Hyman AA, Cobb MH, Kirschner MW (1992) Modulation of the dynamic instability of tubulin assembly by the microtubule-associated protein tau. Mol Biol Cell 3: 1141–1154

Duda JE, Lee VM,Trojanowski JQ (2000) Neuropathology of synuclein aggregates. J Neurosci Res 61: 121–127

Durr A, Stevanin G, Cancel G, Duyckaerts C, Abbas N, Didierjean O, Chneiweiss H, Benomar A, Lyon-Caen O, Julien J, Serdaru M, Penet C, Agid Y, Brice A (1996) Spinocerebellar ataxia 3 and Machado-Joseph disease: clinical, molecular, and neuropathological features. Ann Neurol 39: 490–499

Duyao MP, Auerbach AB, Ryan A, Persichetti F, Barnes GT, McNeil SM, Ge P, Vonsattel JP, Gusella JF, Joyner AL, MacDonald ME (1995) Inactivation of the mouse Huntington's disease gene homolog Hdh. Science 269: 407–410

Feany MB, Bender WW (2000) A Drosophila *model of Parkinson's disease. Nature 404: 394–398*

Fernandez-Funez P, Nino-Rosales ML, de Gouyon B, She WC, Luchak JM, Martinez P, Turiegano E, Benito J, Capovilla M, Skinner PJ, McCall A, Canal I, Orr HT, Zoghbi HY, Botas J (2000) Identification of genes that modify ataxin-1-induced neurodegeneration. Nature 408: 101–106

Gabai VL, Meriin AB, Mosser DD, Caron AW, Rits S, Shifrin VI, Sherman MY (1997) Hsp70 prevents activation of stress kinases. A novel pathway of cellular thermotolerance. J Biol Chem 272: 18033–18037

Giasson BI, Duda JE, Quinn SM, Zhang B, Trojanowski JQ, Lee VM (2002) Neuronal alpha-synucleinopathy with severe movement disorder in mice expressing A53T human alpha-synuclein. Neuron 34: 521–533

Giasson BI, LeeVM (2001) Parkin and the molecular pathways of Parkinson's disease. Neuron 31: 885–888

Goedert M, Spillantini MG, Jakes R, Rutherford D, Crowther RA (1989) Multiple isoforms of human microtubule-associated protein tau: sequences and localization in neurofibrillary tangles of Alzheimer's disease. Neuron 3: 519–526

Gotz J, Chen F, Barmettler R, Nitsch RM (2001a) Tau filament formation in transgenic mice expressing P301L tau. J Biol Chem 276: 529–534

Gotz J, Nitsch RM (2001) Compartmentalized tau hyperphosphorylation and increased levels of kinases in transgenic mice. Neuroreport 12: 2007–2016

Gotz J, Probst A, Spillantini MG, Schafer T, Jakes R, Burki K, Goedert M (1995) Somatodendritic localization and hyperphosphorylation of tau protein in transgenic mice expressing the longest human brain tau isoform. Embo J 14: 1304–1313

Gotz J, Tolnay M, Barmettler R, Chen F, Probst A, Nitsch RM (2001b) Oligodendroglial tau filament formation in transgenic mice expressing G272V tau. Eur J Neurosci13: 2131–2140

Grover A, Houlden H, Baker M, Adamson J, Lewis J, Prihar G, Pickering-Brown S, Duff K, Hutton M (1999) 5' splice site mutations in tau associated with the inherited dementia FTDP-17 affect a stem-loop structure that regulates alternative splicing of exon 10. J Biol Chem 274: 15134–15143

Gusella JF, MacDonald ME (1995) Huntington's disease: CAG genetics expands neurobiology. Curr Opin Neurobiol 5: 656–662.

Hashimoto M, Rockenstein E, Mante M, Mallory M, Masliah E (2001) beta-Synuclein inhibits alpha-synuclein aggregation: a possible role as an anti-parkinsonian factor. Neuron 32: 213–223

Hodgson JG, Agopyan N, Gutekunst CA, Leavitt BR, LePiane F, Singaraja R, Smith DJ, Bissada N, McCutcheon K, Nasir J, Jamot L, Li XJ, Stevens ME, Rosemond E, Roder JC, Phillips AG, Rubin EM, Hersch SM, Hayden MR (1999) A YAC mouse model for Huntington's disease with full-length mutant huntingtin, cytoplasmic toxicity, and selective striatal neurodegeneration. Neuron 23: 181–192

Hutton M, Lendon CL, Rizzu P, Baker M, Froelich S, Houlden H, Pickering-Brown S, Chakraverty S, Isaacs A, Grover A, Hackett J, Adamson J, Lincoln S, Dickson D, Davies P, Petersen RC, Stevens M, de Graaff E, Wauters E, van Baren J, Hillebrand M, Joosse M, Kwon JM, Nowotny P, Kuei Che L, Norton J, Morris JC, Reed LA, Trojanowski J, Basun H, Lannfelt L, Neystat M, Fahn S, Dark F, Tannenberg T, Dodd PR, Hayward N, Kwok JBJ, Schofield PR, Andreadis A, Snowden J, Craufurd D, Neary D, Owen F, Oostra BA, Hardy J, Goate A, van Swieten J, Mann D, Lynch T, Heutink P (1998) Association of missense and 5'-splice-site mutations in tau with the inherited dementia FTDP-17. Nature 393: 702–705

Ikeda H, Yamaguchi M, Sugai S, Aze Y, Narumiya S, Kakizuka A (1996) Expanded polyglutamine in the Machado-Joseph disease protein induces cell death in vitro and in vivo. Nature Genet 13: 196–202

Imai Y, Soda M, Inoue H, Hattori N, Mizuno Y, Takahashi R (2001) An unfolded putative transmembrane polypeptide, which can lead to endoplasmic reticulum stress, is a substrate of Parkin. Cell 105: 891–902

Ishihara T, Hong M, Zhang B, Nakagawa Y, Lee MK, Trojanowski JQ, Lee VM (1999) Age-dependent emergence and progression of a tauopathy in transgenic mice overexpressing the shortest human tau isoform. Neuron 24: 751–762

Jackson GR Salecker I, Dong X, Yao X, Arnheim N, Faber PW, MacDonald ME, Zipursky SL (1998)Polyglutamine-expanded human huntingtin transgenes induce degeneration of Drosophila photoreceptor neurons. Neuron 21: 633–642

Jackson GR, Wiedau-Pazos M, Sang TK, Wagle N, Brown CA, Massachi S, Geschwind DH (2002) Human wild-type tau interacts with wingless pathway components and produces neurofibrillary pathology in Drosophila. Neuron 34: 509–519

Kahle PJ, Neumann M, Ozmen L, Muller V, Jacobsen H, Schindzielorz A, Okochi M, Leimer U, van Der Putten H, Probst A, Kremmer E, Kretzschmar HA, Haass C (2000) Subcellular localization of wild-type and Parkinson's disease- associated mutant alpha -synuclein in human and transgenic mouse brain. J Neurosci 20: 6365–6373

Kawaguchi Y, Okamoto T, Taniwaki M, Aizawa M, Inoue M, Katayama S, Kawakami H, Nakamura S, Nishimura M, Akiguchi I, Kimura J, Narumiya S, Kakizuka A (19940 CAG expansions in a novel gene for Machado-Joseph disease at chromosome 14q32.1. Nature Genet 8: 221–228

Kazantsev A, Walker HA, Slepko N, Bear JE, Preisinger E, Steffan JS, Zhu YZ, Gertler FB, Housman DE, Marsh JL, Thompson LM (2002) A bivalent Huntingtin binding peptide suppresses polyglutamine aggregation and pathogenesis in Drosophila. Nature Genet 30: 367–376

Kazemi-Esfarjani P, Benzer S (2000) Genetic suppression of polyglutamine toxicity in Drosophila. Science 287): 1837–1840

Kitada T, Asakawa S, Hattori N, Matsumine H, Yamamura Y, Minoshima S, Yokochi M, Mizuno Y, Shimizu N (1998) Mutations in the parkin gene cause autosomal recessive juvenile parkinsonism. Nature 392: 605–608

Klement IA, Skinner PJ, Kaytor MD, Yi H, Hersch SM, Clark HB, Zoghbi HY, Orr HT (1998) Ataxin-1 nuclear localization and aggregation: role in polyglutamine- induced disease in SCA1 transgenic mice. Cell 95: 41–53

Kruger R, Kuhn W, Muller T, Woitalla D, Graeber M, Kosel S, Przuntek H, Epplen JT, Schols L, Riess O (1998) Ala30Pro mutation in the gene encoding alpha-synuclein in Parkinson's disease. Nature Genet 18: 106–108

La Spada AR, Wilson EM, Lubahn DB, Harding AE, Fischbeck KH (1991) Androgen receptor gene mutations in X-linked spinal and bulbar muscular atrophy. Nature 352: 77–79

Lee MK, Stirling W, Xu Y, Xu X, Qui D, Mandir AS, Dawson TM, Copeland NG, Jenkins NA, Price DL (2002) Human alpha-synuclein-harboring familial Parkinson's disease-linked Ala- 53 Thr mutation causes neurodegenerative disease with alpha- synuclein aggregation in transgenic mice. Proc Natl Acad Sci USA 99: 8968–8973

Lewis J, McGowan E, Rockwood J, Melrose H, Nacharaju P, Van Slegtenhorst M, Gwinn-Hardy K, Paul Murphy M, Baker M, Yu X, Duff K, Hardy J, Corral A, Lin WL, Yen SH, Dickson DW, Davies P, Hutton M (2000) Neurofibrillary tangles, amyotrophy and progressive motor disturbance in mice expressing mutant (P301L) tau protein. Nature Genet 25: 402–405

Lin X, Antalffy B, Kang D, Orr HT, Zoghbi HY (2000) Polyglutamine expansion down-regulates specific neuronal genes before pathologic changes in SCA1. Nature Neurosci 3: 157–163

Lucas JJ, Hernandez F, Gomez-Ramos P, Moran MA, Hen R, Avila J (2001) Decreased nuclear beta-catenin, tau hyperphosphorylation and neurodegeneration in GSK-3beta conditional transgenic mice. Embo J 20: 27–39

Mangiarini L, Sathasivam K, Seller M, Cozens B, Harper A, Hetherington C, Lawton M, Trottier Y, Lehrach H, Davies SW, Bates GP (1996) Exon 1 of the HD gene with an expanded CAG repeat is sufficient to cause a progressive neurological phenotype in transgenic mice. Cell 87: 493–506

Marsh JL, Walker H, Theisen H, Zhu YZ, Fielder T, Purcell J, Thompson LM (2000) Expanded polyglutamine peptides alone are intrinsically cytotoxic and cause neurodegeneration in Drosophila. Human Mol Genet 9: 13–25

Masliah E, Rockenstein E, Veinbergs I, Mallory M, Hashimoto M, Takeda A, Sagara Y, Sisk A, Mucke L (2000) Dopaminergic loss and inclusion body formation in alpha-synuclein mice: implications for neurodegenerative disorders. Science 287: 1265–1269

McCampbell A, Taylor JP, Taye AA, Robitschek J, Li M, Walcott J, Merry D, Chai Y, Paulson H, Sobue G, Fischbeck KH (2000) CREB-binding protein sequestration by expanded polyglutamine. Human Mol Genet 9: 2197–2202

Moon RT, Bowerman B, Boutros M, Perrimon N (2002) The promise and perils of Wnt signaling through beta-catenin. Science 296: 1644–1646

Mosser DD, Caron AW, Bourget L, Meriin AB, Sherman MY, Morimoto RI, Massie B (2000) The chaperone function of hsp70 is required for protection against stress-induced apoptosis. Mol Cell Biol 20: 7146–7159

Nakamura K, Jeong SY, Uchihara T, Anno M, Nagashima K, Nagashima T, Ikeda S, Tsuji S, Kanazawa I (2001) SCA17, a novel autosomal dominant cerebellar ataxia caused by an expanded polyglutamine in TATA-binding protein. Human Mol Genet 10: 1441–1448

Narhi L, Wood SJ, Steavenson S, Jiang Y, Wu GM, Anafi D, Kaufman SA, Martin F, Sitney K, Denis P, Louis JC, Wypych J, Biere AL, Citron M (1999) Both familial Parkinson's disease mutations accelerate alpha-synuclein aggregation. J Biol Chem 274: 9843–9846

Nucifora FC Jr, Sasaki M, Peters MF, Huang H, Cooper JK, Yamada M, Takahashi H, Tsuji S, Troncoso J, Dawson VL, Dawson TM, Ross CA (2001) Interference by huntingtin and atrophin-1 with cbp-mediated transcription leading to cellular toxicity. Science 291: 2423–2428

Okochi M, Eimer S, Bottcher A, Baumeister R, Romig H, Walter J, Capell A, Steiner H, Haass C (2000) Constitutive phosphorylation of the Parkinson's disease associated alpha-synuclein. J Biol Chem 275: 390–397

Ordway JM, Tallaksen-Greene S, Gutekunst CA, Bernstein EM, Cearley JA, Wiener HW, Dure LS 4th, Lindsey R, Hersch SM, Jope RS, Albin RL, Detloff PJ (1997) Ectopically expressed CAG repeats cause intranuclear inclusions and a progressive late onset neurological phenotype in the mouse. Cell 91: 753–763

Polymeropoulos MH, Lavedan C, Leroy E, Ide SE, Dehejia A, Dutra A, Pike B, Root H, Rubenstein J, Boyer R, Stenroos ES, Chandrasekharappa S, Athanassiadou A, Papapetropoulos T, Johnson WG, Lazzarini AM, Duvoisin RC, Di Iorio G, Golbe LI, Nussbaum RL (1997) Mutation in the alpha-synuclein gene identified in families with Parkinson's disease. Science 276: 2045–2047

Poorkaj P, Bird TD, Wijsman E, Nemens E, Garruto RM, Anderson L, Andreadis A, Wiederholt WC, Raskind M, Schellenberg GD (1998) Tau is a candidate gene for chromosome 17 frontotemporal dementia. Ann Neurol 43: 815–825

Probst A, Gotz J, Wiederhold KH, Tolnay M, Mistl C, Jaton AL, Hong M, Ishihara T, Lee VM, Trojanowski JQ, Jakes R, Crowther RA, Spillantini MG, Burki K, Goedert M (2000) Axonopathy and amyotrophy in mice transgenic for human four-repeat tau protein. Acta Neuropathol (Berl) 99: 469–481

Reddy PS, Housman DE (1997) The complex pathology of trinucleotide repeats. Curr Opin Cell Biol 9: 364–372

Reed LA,Wszolek ZK, Hutton M (2001) Phenotypic correlations in FTDP-17. Neurobiol Aging 22: 89–107

Ross CA (1997) Intranuclear neuronal inclusions: a common pathogenic mechanism for glutamine-repeat neurodegenerative diseases? Neuron 19: 1147–1150

Scherzinger E, Lurz R, Turmaine M, Mangiarini L, Hollenbach B, Hasenbank R, Bates GP, Davies SW, Lehrach H, Wanker EE (1997) Huntingtin-encoded polyglutamine expansions form amyloid-like protein aggregates in vitro and in vivo. Cell 90: 549–558

Sethi KD (2002) Clinical aspects of Parkinson disease. Curr Opin Neurol15: 457–460

Shimohata T, Nakajima T, Yamada M, Uchida C, Onodera O, Naruse S, Kimura T, Koide R, Nozaki K, Sano Y, Ishiguro H, Sakoe K, Ooshima T, Sato A, Ikeuchi T, Oyake M, Sato T, Aoyagi Y, Hozumi I, Nagatsu T, Takiyama Y, Nishizawa M, Goto J, Kanazawa I, Davidson I, Tanese N, Takahashi H, Tsuji S (2000) Expanded polyglutamine stretches interact with TAFII130, interfering with CREB-dependent transcription. Nature Genet 26: 29–36

Shimura H, Hattori N, Kubo S, Mizuno Y, Asakawa S, Minoshima S, Shimizu N, Iwai K, Chiba T, Tanaka K, Suzuki T (2000)Familial Parkinson disease gene product, parkin, is a ubiquitin-protein ligase. Nature Genet 25: 302–305

Shimura H, Schlossmacher MG, Hattori N, Frosch MP, Trockenbacher A, Schneider R, Mizuno Y, Kosik KS, Selkoe DJ (2001) Ubiquitination of a new form of alpha-synuclein by parkin from human brain: implications for Parkinson's disease. Science 293: 263–269

Shubin N, Tabin C, Carroll S (1997) Fossils, genes and the evolution of animal limbs. Nature 388: 639–648

Skinner PJ, Koshy BT, Cummings CJ, Klement IA, Helin K, Servadio A, Zoghbi HY, Orr HT (1997) Ataxin-1 with an expanded glutamine tract alters nuclear matrix- associated structures. Nature 389: 971–974

Spillantini MG, Crowther RA, Jakes R, Hasegawa M, Goedert M (1998a) alpha-Synuclein in filamentous inclusions of Lewy bodies from Parkinson's disease and dementia with lewy bodies. Proc Natl Acad Sci USA 95: 6469–6473

Spillantini MG, Murrell JR, Goedert M, Farlow MR, Klug A, Ghetti B (1998b) Mutation in the tau gene in familial multiple system tauopathy with presenile dementia. Proc Natl Acad Sci USA 95: 7737–7741

Spillantini MG, Schmidt ML, Lee VM, Trojanowski JQ, Jakes R, Goedert M (1997) Alpha-synuclein in Lewy bodies. Nature 388: 839–840

Spittaels K, Van den Haute C, Van Dorpe J, Bruynseels K, Vandezande K, Laenen I, Geerts H, Mercken M, Sciot R, Van Lommel A, Loos R, Van Leuven F (1999) Prominent axonopathy in the brain and spinal cord of transgenic mice overexpressing four-repeat human tau protein. Am J Pathol 155: 2153–2165

Steffan JS, Bodai L, Pallos J, Poelman M, McCampbell A, Apostol BL, Kazantsev A, Schmidt E, Zhu YZ, Greenwald M, Kurokawa R, Housman DE, Jackson GR, Marsh JL, Thompson LM (2001) Histone deacetylase inhibitors arrest polyglutamine-dependent neurodegeneration in Drosophila. Nature 413: 739–743

Steffan JS, Kazantsev A, Spasic-Boskovic O, Greenwald M, Zhu YZ, Gohler H, Wanker EE, Bates GP, Housman DE, Thompson LM (2000) The Huntington's disease protein interacts with p53 and CREB-binding protein and represses transcription. Proc Natl Acad Sci USA 97: 6763–6768

Tanemura K, Akagi T, Murayama M, Kikuchi N, Murayama O, Hashikawa T, Yoshiike Y, Park JM, Matsuda K, Nakao S, Sun X, Sato S, Yamaguchi H, Takashima A (2001) Formation of filamentous tau aggregations in transgenic mice expressing V337M human tau. Neurobiol Dis 8: 1036–1045

Trottier Y, Devys D, Imbert G, Saudou F, An I, Lutz Y, Weber C, Agid Y, Hirsch EC, Mandel JL (1995) Cellular localization of the Huntington's disease protein and discrimination of the normal and mutated form. Nature Genet 10: 104–110

van der Putten H, Wiederhold KH, Probst A, Barbieri S, Mistl C, Danner S, Kauffmann S, Hofele K, Spooren WP, Ruegg MA, Lin S, Caroni P, Sommer B, Tolnay M, Bilbe G (2000) Neuropathology in mice expressing human alpha-synuclein. J Neurosci 20: 6021–6029

Warrick JM, Chan HY, Gray-Board GL, Chai Y, Paulson HL, Bonini NM (1999) Suppression of polyglutamine-mediated neurodegeneration in Drosophila by the molecular chaperone HSP70. Nature Genet 23: 425–428

Warrick JM, Paulson HL, Gray-Board GL, Bui QT, Fischbeck KH, Pittman RN, Bonini NM (1998) Expanded polyglutamine protein forms nuclear inclusions and causes neural degeneration in Drosophila. Cell 93: 939–949

Watase K, Weeber EJ, Xu B, Antalffy B, Yuva-Paylor L, Hashimoto K, Kano M, Atkinson R, Sun Y, Armstrong DL, Sweatt JD, Orr HT, Paylor R, Zoghbi HY (2002) A long CAG repeat in the mouse Sca1 locus replicates SCA1 features and reveals the impact of protein solubility on selective neurodegeneration. Neuron 34: 905–919

Wexler NS, Young AB, Tanzi RE, Travers H, Starosta-Rubinstein S, Penney JB, Snodgrass SR, Shoulson I, Gomez F, Ramos Arroyo MA, Penchaszadeh GK, Moreno H, Gibbons K, Faryniarz A, Hobbs W, Anderson MA, Bonilla E, Conneally PM, Gusella JF (1987) Homozygotes for Huntington's disease. Nature 326: 194–197

Wittmann CW, Wszolek MF, Shulman JM, Salvaterra PM, Lewis J, Hutton M, Feany MB (2001) Tauopathy in Drosophila: neurodegeneration without neurofibrillary tangles. Science. 293: 711–714

Yamamoto A, Lucas JJ, Hen R (2000) Reversal of neuropathology and motor dysfunction in a conditional model of Huntington's disease. Cell 101: 57–66

Yin JC, Tully T (1996) CREB and the formation of long-term memory. Curr Opin Neurobiol 6: 264–268

Yoshida H, Ihara Y (1993) Tau in paired helical filaments is functionally distinct from fetal tau: assembly incompetence of paired helical filament-tau. J Neurochem 61: 1183–1186

Yvert G, Lindenberg KS, Devys D, Helmlinger D, Landwehrmeyer GB, Mandel JL (2001) SCA7 mouse models show selective stabilization of mutant ataxin-7 and similar cellular responses in different neuronal cell types. Human Mol Genet 10: 1679–1692

Zeitlin S, Liu JP, Chapman DL, Papaioannou VE, Efstratiadis A (1995) Increased apoptosis and early embryonic lethality in mice nullizygous for the Huntington's disease gene homologue. Nature Genet 11: 155–163

Zhang S, Xu L, Lee J, Xu T (2002) Drosophila atrophin homolog functions as a transcriptional corepressor in multiple developmental processes. Cell 108: 45–56

Zhang Y, Gao J, Chung KK, Huang H, Dawson VL, Dawson TM (2000) Parkin functions as an E2-dependent ubiquitin- protein ligase and promotes the degradation of the synaptic vesicle-associated protein, CDCrel-1. Proc Natl Acad Sci USA 97: 13354–13359

Zhou H, Li SH, Li XJ (2001)Chaperone suppression of cellular toxicity of huntingtin is independent of polyglutamine aggregation. J Biol Chem 276: 48417–48424

Zipursky SL, Rubin GM (1994) Determination of neuronal cell fate: lessons from the R7 neuron of Drosophila. Annu Rev Neurosci 17:373–397

Zoghbi HY,.Orr HT (1995) Spinocerebellar ataxia type 1. Semin Cell Biol 6: 29–35

# Human behavioral genomics

*R. Plomin*[1]

## Summary

We are rapidly approaching a postgenomic era in which we will know the entire genome sequence and, most importantly to human behavioral genetics, all the variations in the genome sequence. These DNA polymorphisms are the source of hereditary influence and will revolutionize human behavioral genetics, especially links between genes, brain and behavior.

Attention is now focused on finding genes for complex quantitative traits influenced by multiple genes (called quantitative trait loci, QTLs) as well as environmental factors. Behavioral disorders and dimensions are the most complex of all quantitative traits. Progress in identifying genes for complex traits has been slower than expected in part because research has lacked statistical power to detect genes of small effect size. Research designs are needed that can break the 1% QTL barrier.

Much has been learned and remains to be learned using quantitative genetic techniques that can disentangle genetic and environmental influences. In addition to demonstrating that genetic influence is substantial for most behavioral disorders and dimensions, quantitative genetic techniques can be used to assess genetic and environmental influences on covariance between behavioral traits or between behavioral and biological traits (multivariate genetics), to chart genetic change as well as continuity during development (developmental genetics), and to investigate interactions and correlations between genes and environment (environmental genetics). This research can chart the course for molecular genetic attempts to identify QTLs.

The greatest impact of the postgenomic era on human behavioral genetics research will come after specific genes have been identified that contribute to the ubiquitous heritability of behavioral traits. The future of genetic research lies in moving from finding genes (genomics) to finding out how genes work (functional genomics). Functional genomics is usually considered in terms of bottom-up molecular biology at the cellular level of analysis. However, a top-down behavioral level of analysis (behavioral genomics) might pay off more quickly in predicting, diagnosing, treating and preventing problems. Bottom-up and top-

---

[1] MRC Research Professor and Deputy Director Social, Genetic and Developmental Psychiatry Research Centre, Institute of Psychiatry, Kings College London, BOX PO80, London SE5 8AF, United Kingdom

Mallet/Christen
Neurosciencess at the Postgenomic Era
© Springer-Verlag Berlin Heidelberg 2003

down levels of analysis of gene-behavior pathways will eventually meet in the brain.

## Introduction

The 20th century began with the re-discovery of Mendels laws of heredity. The word "genetics" was first coined in 1903. Fifty years later the double helix structure of DNA was discovered. The genetic code was cracked in 1966 – the four-letter alphabet (G, A, T, C) of DNA is read as three-letter words that code for the 20 amino acids that are the building blocks of proteins. The crowning glory of the century and the beginning of the new millennium is the Human Genome Project, which has provided a working draft of the sequence of the three billion letters of DNA in the human genome.

We are now at the threshold of a postgenomic world in which all genes and all DNA variation will be known. The major beneficiary of this new world will be research on complex traits that are influenced by multiple genes as well as multiple environmental influences. This is where behavior comes in. Behavioral dimensions and disorders are the most complex traits of all and are already becoming a focus for DNA analysis. The genetic analysis of behavior includes both quantitative genetic (twin and adoption studies) and molecular genetic (DNA) strategies, the two worlds of genetics that are coming together in the study of complex traits such as behavioral dimensions and disorders.

The goal of this chapter is to discuss where we are and where we are going in genetic research on behavior, especially in relation to brain mechanisms that mediate genetic effects on behavior. Nearly a decade ago, an edited volume summarized human behavioral genetic research (Plomin and McClearn 1993). At that time, quantitative genetic research, such as twin and adoption studies of human behavior and strain and selection studies in other species, had made the case that nearly all behavioral dimensions and disorders show at least moderate genetic influence. What was new was that quantitative genetic research was going beyond estimating heritability to ask questions about developmental change and continuity in genetic influence, about multivariate issues such as co-morbidity and heterogeneity, and about the interface between genes and environment, topics discussed later in the present chapter. A few chapters in the 1993 volume mentioned the potential of molecular genetics to identify some of the many genes that are responsible for the ubiquitous heritability of behavioral dimensions and disorders. However, few confirmed linkages and associations for complex behavioral traits were known at that time. Although a few chapters in the 1993 volume mentioned the potential of molecular genetics, the only chapter to have a section on molecular genetics was a chapter on reading disability (DeFries and Gillis 1993), because the 6p21 linkage with reading disability was just beginning to emerge at that time. It was the first QTL linkage found for a human behavioral disorder (Cardon et al. 1994) and has subsequently been replicated in several studies (Willcutt et al. 2003). However, few confirmed linkages and associations for complex behavioral traits were known at that time.

As genes associated with behavior are identified, the next step is to understand the mechanisms by which genes have their effect (functional genomics). The phrase "behavioral genomics" refers to understanding how genes affect behavior at the level of functioning of the whole organism. The purpose of this chapter is to explain why I believe that the future of behavioral genetics in a post-genomic era is behavioral genomics.

## The Very Standard Deviation

It is important to begin with a discussion of the different perspectives or levels of analysis used to investigate behavior because so much follows conceptually as well as methodologically from these differences in perspective (see Fig. 1.) Research in behavioral genetics has focused on within-species inter-individual differences – for example, why some children are reading disabled and others are not. In contrast, many areas of psychology and neuroscience seldom mention individual differences and concentrate instead on species-universal or species-typical (normative) phenomena. For example, leading textbooks in cognitive neuroscience (e.g., Gazzaniga 2000; Thompson 2000) do not mention individual differences. Neuroscience has focused on understanding how the brain works on average – for example, which bits of the brain light up under neuroimaging for particular tasks. Until now, genetics has entered neuroscience primarily in relation to gene targeting in mice in which mutations are induced that interfere with

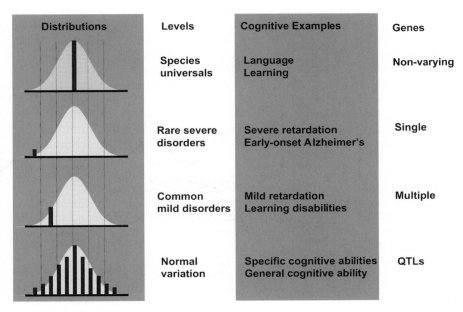

**Fig. 1.** Levels of analysis.

normal brain processes. In humans, rare single-gene mutations have been the center of attention. This approach tends to treat all members of the species as if they were genetically the same except for a few rogue mutations that disrupt normal processes. In this sense, the species-typical perspective of neuroscience assumes that mental illness is a broken brain. In contrast, the individual differences perspective considers variation as normal – the very standard deviation. Common behavioral disorders such as learning disabilities or mental illness are thought to be the quantitative extreme of the normal distribution.

Perspectives are not right or wrong – just more or less useful for particular purposes. However, the species-typical perspective and the individual differences perspective can arrive at different answers because they ask different questions. The distinction between the two perspectives is in essence the difference between means and variances. There is no necessary connection between means and variances, either descriptively or etiologically. Despite its name, analysis of variance, the most widely used statistical analysis in the life sciences, is actually an analysis of mean effects in which individual differences are literally called the "error term". Instead of treating differences between individuals as error and averaging individuals across groups as in analysis of variance, individual differences research focuses on these inter-individual differences. Variation is distributed continuously, often in the shape of the familiar bell curve, and is indexed by variance (the sum of squared deviations from the mean) or the square of variance, which is called the standard deviation.

The two perspectives also differ methodologically. Most species-typical research is experimental in the sense that subjects are randomly assigned to conditions that consist of manipulating something such as genes, lesions, drugs, or tasks. The dependent variable is the average effect of the manipulation on outcome measures such as single-cell recordings of synaptic plasticity, activation of brain regions assessed by neuroimaging, or performance on cognitive tests. Such experiments ask whether manipulations *can* have an effect on average in a species. For example, a gene knock-out study investigates whether an experimental group of mice who inherit a gene that has been made dysfunctional differs, for example, in learning or memory from a control group with a normal copy of the gene. A less obvious example can be seen in recent experimental research that manipulated tasks and found that average blood flow assessed by PET in the human species is greater in the prefrontal cortex for high-intelligence tasks than for low-intelligence tasks (Duncan et al. 2000).

In contrast, rather than creating differences between experimental and control groups through manipulations, the individual differences perspective focuses on naturally occurring differences between individuals. One of the factors that make individuals different is genetics. The individual differences perspective is the foundation for quantitative genetics, which focuses on naturally occurring genetic variation, the stuff of heredity. Although 99.9% of the human DNA sequence is identical for all human beings, the 0.1% that differs – three million base pairs – is ultimately responsible for the ubiquitous genetic influence found for complex traits including behavioral dimensions and disorders (Plomin et al. 2001b). Individual differences research is correlational in the sense that it investigates factors that *do* have an effect in the world outside the laboratory. Continuing with the previous examples, in contrast to the knock-out approach, an individual differences approach asks whether naturally occurring genetic variation in mice is associated with individual differences in mouse learning and memory. [The answer to this question is yes (Wehner and Balogh, 2003).] The brain is like the engine of an automobile, with thousands of parts that need to work in order for the engine to run properly. In trying to diagnose the defect in a faulty engine, we may test the effect of removing any of these parts. The engine may not run properly as a consequence but this part is unlikely to be the reason why a particular automobile is not running properly. In the same way, genes can be knocked out and be shown to have major effects on learning and memory but this does not imply that the gene has anything to do with the naturally occurring genetic variation that is responsible for hereditary transmission of individual differences in performance on learning and memory tasks. The PET experiment that compared average performance on high- and low-intelligence tasks could employ an individual differences perspective by comparing cortical blood flow in high and low intelligence individuals rather than comparing average performance on tasks. An example of individual differences research is a recent twin study that found very high heritability for individual differences in MRI-assessed gray matter density in the frontal cortex and showed that these individual differences covary with individual differences in intelligence (Thompson et al. 2002).

Other perspectives lie in between these two extremes of species universals and normal variation. The effects of rare severe disorders caused by single-gene defects can be dramatic. For example, mutations in the gene that codes for the enzyme phenylalanine hydroxylase, if untreated, cause phenylketonuria (PKU), which is associated with a severe form of mental retardation. This inherited condition occurs in one in 10,000 births. At least 100 other rare, single-gene disorders include mental retardation as a symptom of the disorder (Wahlström 1990). Such rare, single-gene disorders are viewed from the species-universals perspective as aberrations from the species type, exceptions to the species rule. However, common disorders – such as mild mental retardation and learning disabilities – seldom show any sign of single-gene effects and appear to be caused by multiple genes as well as by multiple environmental factors. Indeed, quantitative genetic research suggests that such common disorders are usually the quantitative extreme of the same genes responsible for variation throughout the distribution. Genes in such multiple-gene (polygenic) systems are called quantitative trait loci (QTLs) because they are likely to result in dimensions (quantitative continua) rather than disorders (qualitative dichotomies). In other words, in terms of the genetic etiology of common disorders, there may be no disorders, just dimensions. That is, other than simple and rare single-gene or chromosomal disorders, mental illness may represent the extreme of normal variation.

In summary, the individual differences perspective views variation as normal and distributed continuously; common disorders are viewed as the extremes of these continuous distributions. Perspectives are not right or wrong, but they are different and the proper interpretation of genetic research depends on understanding these differences. The perspectives are complementary, and a full understanding of behavior in a post-genomic era requires integration across all levels of analysis.

## The Two Worlds of Genetics

The species-universal perspective and the individual differences perspective contributed to the drifting apart of the two worlds of quantitative genetics and molecular genetics since their origins a century ago. Quantitative genetics focuses on naturally occurring individual differences in complex phenotypes that are commercially important in agriculture and animal husbandry or socially important in the human species. Genetic tools were developed to assess the extent to which such traits were inherited. In nonhuman animals, these tools included experimental crosses, inbred strains and artificial selection. In the human species, twin and adoption methods were developed in the 1920s. The essence of the theoretical foundation for quantitative genetics is the extension of Mendel's laws of single-gene inheritance to complex multifactorial traits influenced by multiple genes as well as multiple environmental factors (Fisher 1918). If multiple genes affect a trait, the trait is likely to be distributed quantitatively as each gene adds its effects to the mix. The goal of quantitative genetics is to decompose observed variation in a trait and covariation between traits into genetic and environmental sources of variation. The conclusion has gradually emerged from quantitative genetic research that genetic factors contribute to nearly all complex traits, including behavior. However, the polygenic nature of complex traits makes seem unlikely that we will ever identify specific genes responsible for the inheritance of complex traits.

In contrast to the focus of quantitative genetics on naturally occurring sources of variation in complex traits, the origins of molecular genetics can be traced back to species-universal research early in the century that created new genetic variation by mutating genes chemically or through X-irradiation. This experimental orientation continues today with large-scale mutation screening research programs and targeted mutations such as deleting key DNA sequences that prevent a particular gene from being transcribed ("knock-outs") or alter its regulation in order to underexpress or overexpress the gene (Capecchi 1994). The focus

Fig. 2. Two worlds of genetics.

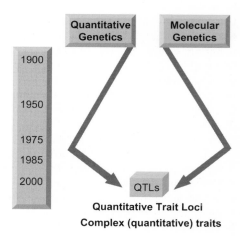

Quantitative Trait Loci
Complex (quantitative) traits

here is on the gene rather than the phenotype and on creating new genetic variation rather than studying the phenotypic effects of naturally occurring genetic variation. As a result, these approaches concentrate on single genes rather than multiple genes and on simple phenotypes to assess the average effects of mutations on the species rather than individual differences in complex phenotypes that are of interest in their own right.

These conceptual and methodological differences caused the two worlds of genetics to drift apart until the 1980s, when a new generation of DNA markers made it possible to identify genes for complex quantitative traits influenced by multiple genes (QTLs) as well as by multiple environmental factors. Until the 1980s, only a few genetic markers were available that were the products of single genes, such as the red blood cell proteins that define the blood groups. The new genetic markers were polymorphisms in DNA itself rather than in a gene product. Because millions of DNA base sequences are polymorphic, millions of such DNA markers became potentially available to hunt for QTLs responsible for the inheritance of complex traits.

Unlike single-gene effects that are necessary and sufficient for the development of a disorder, QTLs contribute incrementally, analogous to probabilistic risk factors (see Fig. 3.) The resulting trait is distributed quantitatively as a dimension rather than qualitatively as a disorder; this was the essence of Fisher's classic 1918 paper on quantitative genetics (Fisher 1918). The QTL perspective is the molecular genetic version of the quantitative genetic perspective, which assumes that genetic variance on complex traits is due to many genes of varying effect size. As mentioned earlier, from a QTL perspective, common disorders may be just the extremes of quantitative traits caused by the same genetic and environmental factors responsible for variation throughout the dimension. In other words, the QTL perspective predicts that genes found to be associated with complex disorders will also be associated with normal variation on the same dimension and vice versa (Deater-Deckard et al. 1997; Plomin et al. 1994).

In contrast to the species-universal perspective, in which we are all thought to be the same genetically except for a few rogue mutations that lead to disorders, the QTL perspective suggests that genetic variation is normal. Many genes affect most complex traits and, together with environmental variation, these QTLs are responsible for normal variation as well as for the abnormal extremes of these quantitative traits. This QTL perspective has implications for thinking about mental illness because it blurs the etiological boundaries between the normal and the abnormal. That is, we all have many alleles that contribute to mental illness but some of us are unluckier in the hand that we drew at conception from our parents genomes. A more subtle conceptual advantage of a QTL perspective is that it frees us to think about both ends of the normal distribution:the positive end as well as the problem end, abilities as well as disabilities, and resilience as well as vulnerability. It has been proposed that we move away from an exclusive focus on pathology toward considering positive traits that improve the quality of life and perhaps prevent pathology (Seligman and Csikszentmihalyi 2000).

The initial goal of QTL research is not to find *the* gene for a complex trait but rather those genes that make contributions of varying effect sizes to the variance of the trait. Perhaps one gene will be found that accounts for 5% of the variance,

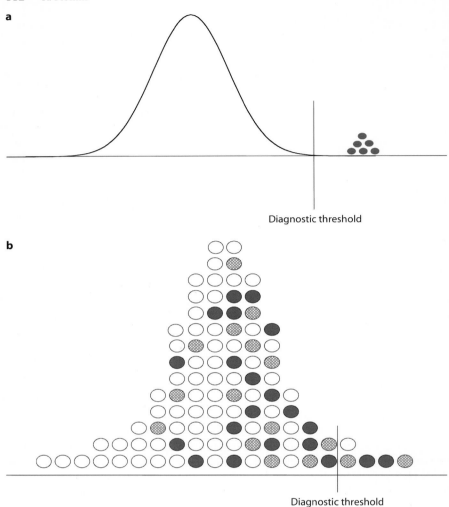

**Fig. 3. a** Single genes are necessary and sufficient to develop a disorder. **b** QTLs are probabilistic risk factors neither necessary nor sufficient to develop a disorder. Dark dots represent individuals who have an allele associated with a disorder. The cross-hatched dots in **b** represent the allele of a second locus that is associated with a disorder. A QTL perspective implies that common disorders may be the quantitative extreme of the same genetic factors responsible for variation throughout the normal distribution.

five other genes might each account for 2% of the variance, and 10 other genes might each account for 1% of the variance. If the effects of these QTLs are independent, together these QTLs would account for 25% of the traits variance. All of the genes that contribute to the heritability of a complex trait are unlikely to be identified because some of their effects may be too small to detect. The problem is that we do not know the distribution of effect sizes of QTLs for any complex trait in plant, animal or human species. Not long ago, a 10% effect size was thought to be small, at least from the single-gene perspective in which the effect size was essentially 100%. However, for behavioral disorders and dimensions, a 10% effect size may turn out to be a very large effect. Perhaps a 1% effect size will be a relatively large effect.

If effect sizes are most often smaller than 1%, this would explain the slow progress to date in identifying genes associated with behavior, because research so far is woefully underpowered to detect and replicate QTLs of small effect size (Cardon and Bell 2001). Although the tremendous advances during the past few years from the Human Genome Project warrant optimism for the future, progress in identifying genes associated with behavioral traits has clearly been slower than expected. Using schizophrenia and bipolar affective disorder as examples, no clear-cut associations with schizophrenia and bipolar affective disorder have been identified despite a huge effort during the past decade, although there are several promising leads (Baron 2001). Part of the reason for this slow progress may be that, because these were the first areas in which molecular genetic approaches were applied, they happened at a time in the 1980s when large pedigree linkage designs were in vogue. It is now generally accepted that such designs are only able to detect genes of major effect size.

Recent research has been more successful for finding QTLs for complex traits because different designs have been employed that can detect genes of much smaller effect size, for example, for reading disability (Willcutt et al. 2002), anxiety (Lesch 2003), and hyperactivity (Thapar 2003). QTL linkage designs use many small families (usually siblings) rather than a few large families and they are able to detect genes that account for about 10% of the variance. Association studies such as case-control comparisons make it possible to detect genes that account for much smaller amounts of variance (Risch 2000; Risch and Merikangas 1996). A daunting target for molecular genetic research on complex traits such as behavior is to design research powerful enough to detect QTLs that account for 1% of the variance while providing protection against false positive results in genome scans using thousands of markers. In order to break the 1% QTL barrier (which no study has yet done), samples of many thousands of individuals are needed for research on disorders (comparing cases and controls) and on dimensions (assessing individual differences in a representative sample).

It should be noted that DNA variation has a unique causal status in explaining behavior. When behavior is correlated with anything else, the old adage applies that correlation does not imply causation. For example, when parenting is shown to be correlated with children's behavioral outcomes, this does not imply that the parenting caused the outcome environmentally. Behavioral genetic research has shown that parenting behavior to some extent reflects genetic influences on children's behavior, as mentioned in the following section. When it comes to interpreting correlations between biology and behavior, such correlations are often mistakenly interpreted as if biology causes behavior. For example, correlations between neurotransmitter physiology and behavior or between neuroimaging indices of brain activation and behavior are often interpreted as if brain differences cause behavioral differences. However, these correlations do not necessarily imply causation, because behavioral differences can cause brain differences. In contrast, in the case of correlations between DNA variants and behavior, the behavior of individuals does not change their genome. Expression of genes can be altered but the DNA sequence itself does not change. For this reason, correlations between DNA differences and behavioral differences can be interpreted causally: DNA differences can cause the behavioral differences but not the other way around.

## Quantitative Genetics: Beyond Heritability

Because quantitative genetics considers the net effect of all genes in the genome on a particular trait, it can be viewed as a genomic approach although it does not specify the specific genes that affect the trait. Quantitative genetic research can chart the course for molecular genetic research, not only in terms of documenting a trait's heritability but also in going beyond estimating heritability to ask multivariate, developmental and environmental questions, as described below.

The main success of quantitative genetics to date has been to provide convincing evidence for the nearly ubiquitous importance of genetic influence on behavioral disorders and dimensions. Thanks to quantitative genetics, the behavioral sciences have emerged from an era of strict environmental explanations for differences in behavior to a more balanced view that recognizes the importance of nature (genetics) as well as nurture (environment). This shift occurred first for behavioral disorders, including rare disorders such as autism (0.001 incidence), more common disorders such as schizophrenia (0.01), and very common disorders such as reading disability (0.05). More recently it has become increasingly accepted that genetic variation contributes importantly to differences among individuals in the normal range of variability as well as for abnormal behavior (Plomin et al. 2001b). The most well-studied domains of behavior are cognitive abilities and disabilities, psychopharmacology, personality and psychopathology. A growth area in neuroscience is the application of quantitative genetics, as seen for example in the recent twin study mentioned earlier that found high heritability of MRI-assessed gray matter density in several brain regions (Thompson et al. 2002).

Because of the rarity of newborn adoptees, the twin method has provided the bulk of human quantitative genetic research – about 1% of all births are twins. The twin method is a natural experiment that compares similarity of identical and fraternal twins. Possible problems with the twin method have been examined, but it is generally a robust method and yields results similar to the other major method, the adoption method (Plomin et al. 2001b).

Figure 4 summarizes twin results for 1) common medical disorders, 2) common psychiatric disorders, and 3) psychological dimensions. Identical (monozygotic, MZ) twins are generally more similar than fraternal (dizygotic, DZ) twins, suggesting that genetic influence is important for most disorders and dimensions. This overview illustrates several points. First, intuitions about what is and what is not heritable are not worth much. For example, who would have guessed that ulcers and epilepsy are much more highly heritable than Parkinson's disease and breast cancer (Fig. 4a) A second point is that twin data provide the best available evidence for the importance of the environment because in no case does genetic influence account for all the variance. Because identical twins are genetically identical, like clones, no genetic explanation can account for differences within pairs of identical twins. For example, the finding that identical twin sisters are only about 15% concordant for breast cancer implies that environmental factors are primarily responsible for breast cancer. Although genes responsible for some early-onset, severe forms of breast cancer have been found despite the low heritability of breast cancer, it should be easier to find genes for disorders that are more highly heritable. A third point is that behavioral disorders (Fig. 4b) tend to be more heritable than medical disorders. Why this is so is not known, but it may be that. in the flow from genes to brain to behavior, behavior is highly multifactorial in the sense of being so far downstream and thus picks up genetic influences from many sources. Regardless of the reason, the high heritability of behavioral disorders suggests that they are good targets for molecular genetic attempts to identify genes. Finally, genetic influence is not restricted to disorders – heritability is substantial for normal behavioral variation, such as personality and cognitive abilities (Fig. 4c). For example, general cognitive ability (sometimes called intelligence) in dozens of twin studies with a total of more than 10,000 twin pairs shows MZ and DZ correlations of about 0.85 and 0.60, respectively. This pattern of twin correlations yields a heritability estimate of about 50%, indicating that about half of the variance of general cognitive ability can be accounted for by genetic facors. Because of its special relevance to neuroscience, cognitive ability will be used below as an example of research that goes beyond merely estimating heritability.

Although twin and adoption studies of individual differences in previously neglected domains such as neuroscience will yield important new insights into their etiology, asking only how much genetic factors influence such domains does not fully capitalize on the strengths of quantitative genetic analysis. New quantitative genetic techniques make it possible to go beyond the rudimentary question of heritability to investigate multivariate, developmental and environmental issues. These techniques chart the course for molecular genetic studies by identifying the most heritable components and constellations of disorders as they develop and as genetic vulnerabilities correlate and interact with the environment. In this sense, quantitative genetics will be even more important in a postgenomic era.

**Multivariate genetics.** Multivariate genetic analysis focuses on the etiology of the covariance between traits and can be used to investigate such issues as comorbidity and heterogeneity, links between behavior and biology, and links between the normal and abnormal. Using cognitive abilities as an example, tests of diverse cognitive abilities, such as verbal, spatial and memory, intercorrelate about 0.30 on average in a meta-analysis of 322 studies (Carroll 1993). A technique called factor analysis indicates that a general factor (first unrotated principal component) accounts for about 40% of the total variance of tests of diverse cognitive abilities. This general factor representing what such tests have in common is what is meant by general cognitive ability (Plomin 1999). Multivariate

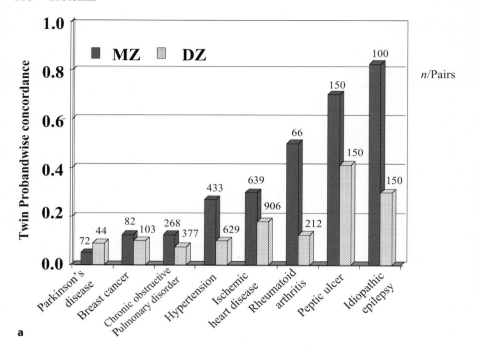

a

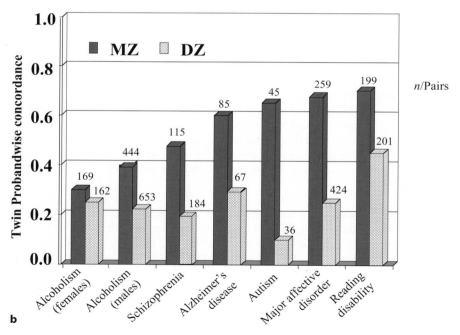

b

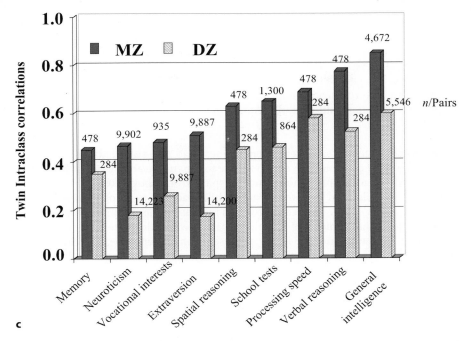

**Fig.4.** Identical (monozygotic, MZ) and fraternal (dizygotic, DZ) twin resemblances for **(a)** common medical disorders, **(b)** psychiatric disorders, and **(c)** psychological dimensions. The number above each bar is the number of twin pairs. (adapted from Plomin et al. 1994.)

genetic analysis of the covariance among cognitive tests yields a surprising finding with far-reaching ramifications for neuroscience (Plomin and Spinath 2002). Although a general factor accounts for about 40% of the phenotypic variance of cognitive tests, a general genetic factor accounts for nearly all of the genetic variance of cognitive tests. In others words, if a gene were identified that is associated with a particular psychometrically assessed cognitive ability, the same gene would be expected to be associated with other cognitive abilities as well. In this way, multivariate genetic analysis suggests that there is much genetic overlap (comorbidity) among cognitive processes.

Multivariate genetic analysis can also be used to investigate genetic links between behavior and biology. For example, in terms of general cognitive ability, multivariate genetic research has begun to consider the brain processes responsible for genetic overlap among cognitive abilities. It appears that this genetic overlap occurs not just for traditional tests of cognitive abilities but also for reaction time measures of elementary cognitive processes and perhaps even for physical and physiological brain measures (Plomin and Spinath 2002).

Finally, a type of multivariate genetic analysis can be used to explore links between the normal and abnormal. This research suggests that both ends of the normal distribution of general cognitive ability – the low end of mild mental retardation and the high end of giftedness – are due to the same genetic and environmental influences that operate throughout the distribution, which supports the QTL perspective described earlier. An exception is severe mental retardation, which is generally not heritable in that it is caused by environmental trauma such as birth problems or genetic trauma that is not inherited, such as chromosomal abnormalities and spontaneous genetic mutations (Plomin et al. 2001b).

**Developmental genetics.** Developmental questions can be asked about genetic change as well as continuity. A surprise here for cognitive ability is that heritability increases steadily throughout the life span, from about 20% in infancy to about 40% in childhood to about 60% in adulthood and perhaps to 80% in later life (McGue et al. 1993). The reason for this increase in heritability is not known, but it suggests that molecular genetic research would do better to study older individuals whose phenotype better represents their genotype. Developmental analysis of longitudinal data suggests some interesting developmental changes from age to age. For example, new genetic influences appear to emerge during the early school years, an age at which all theories of cognitive development posit major change. However, most genetic influence contributes to continuity rather than to age-to-age change (Plomin et al. 2001b).

**Environmental genetics.** A third direction for research that goes beyond merely estimating heritabilities is to study the interface between nature and nurture. The present chapter's focus on genetics is not intended to denigrate the importance of environmental influences or to imply biological determinism. In many ways, it is more difficult to study the environment than genes. There is no Human Environome Project – indeed, environmental research is premendelian in the sense that the laws of environmental transmission and even the units of transmission are unknown. Although this chapter has focused on genetics, it should be emphasized that quantitative genetics, unlike molecular genetics, is at least as informative about nurture as it is about nature. As noted earlier, quantitative genetic research provides the best available evidence for the importance of the environment, in that the heritability of complex traits is seldom greater than 50%. In addition, two of the most important findings about environmental influences on behavior have come from genetic research.

The first finding is that, contrary to socialization theories from Freud onward, environmental influences operate to make children growing up in the same family as different as children growing up in different families, which is called nonshared environment (reviewed by Plomin et al. 2001a). General cognitive ability shows an interesting twist on the finding that most environmental influences are of the nonshared variety. For general cognitive ability, shared environmental influences are important in childhood, accounting for about a quarter of the total variance. However, shared environmental influence diminishes to negligible importance after adolescence as children make their own way in the world outside the family. Thus, even for cognitive ability, nonshared environmental influences are most important in the long run.

The second finding, called the nature of nurture, is that genetic factors influence the way we experience our environments (reviewed by Plomin 1994). This is an example of genotype-environment correlation, in which genetic disposition is correlated with environmental exposure. Genotype-environment correlation explains the puzzling finding that most measures of the environment used in behavioral research show genetic influence. It also explains why associations between environmental measures and behavioral outcome measures are often substantially mediated genetically. Using cognitive ability again as an example, observational and interview measures of family environment show substantial genetic influence. Moreover, correlations between measures of family environment and childrens cognitive development are substantially mediated genetically. These findings suggest that, in addition to parents contributing environmentally to their childrens cognitive development, parents are also responding to genetic factors influencing their children's cognitive ability.

Another type of interface between nature and nurture is genotype-environment interaction, which involves genetic sensitivity to the environment (Kendler and Eaves 1986). One form of interaction occurs when heritability differs as a function of environment. For cognitive ability, some research suggests a genotype-environment interaction, in that heritability appears to be lower for children in poorer families, perhaps because poorer backgrounds swamp genetic potential (Rowe et al. 1999). In general, however, it has been much more difficult to find replicable examples of genotype-environment interaction than genotype-environment correlation.

In summary, quantitative genetics can go beyond merely estimating heritability to ask more informative multivariate, developmental and environmental questions. The answers to these questions can inform molecular genetic research. When specific genes are found that are associated with behavioral traits, it will be possible to address these same sorts of questions with much greater precision (Table 1).

## Molecular Genomics

Although the working draft of the human genome sequence published in February 2001 was a stunning achievement, much work still needs to be done to complete the sequence and to identify all of the genes. For behavioral genetics, the most important next step is the identification of the DNA sequences that make us different from each other. There is no single human genome sequence – we each have a unique genome. Most of the DNA letters are the same for all human genomes, and many of these are the same for other primates and other mammals and even insects. Nevertheless, about one in every thousand DNA letters differs among us, around three million DNA variations in total, enough to make us each differ for almost every gene. Most of these DNA differences involve a substitution of a single base pair, called single nucleotide polymorphisms (SNPs, pronounced "snips"). These DNA differences are responsible for the widespread heritability of behavioral disorders and dimensions. That is, when we say that a trait is heritable, we mean that variations in DNA exist that cause differences in behav-

**Table 1.** Examples of quantitative genetic findings and the questions they raise for behavioral genomics.

| Quantitative genetics findings | Behavioral genomics questions |
| --- | --- |
| *Heritability/QTLs* | |
| Heritability established for most traits | Can QTLs be identified? |
| *Developmental issues* | |
| Cross-sectional: | |
| Heritability tends to increase with age | Does the strength of QTL associations increase with age? |
| Longitudinal: | |
| From age to age, genetic influence shows much continuity and some change | To what extent do QTLs early in life predict later behavior? |
| *Multivariate issues* | |
| Heterogeneity and comorbidity: | |
| Some disorders consist of genetically different components (heterogeneity); some supposedly different disorders overlap genetically (comorbidity) | Can QTLs create a genetically based nosology? |
| Links between normal and "abnormal": | |
| Common disorders are generally the quantitative extreme of the same genetic factors responsible for heritability throughout the distribution | Are QTLs that are associated with disorders the same QTLs associated with normal variation? |
| *Environmental issues* | |
| GE interaction: | |
| Heritability sometimes differs as a function of environnement | Do QTL associations differ as a function of environment? |
| GE correlation: | |
| Associations between environmental measures and behavior are often mediated genetically | To what extent do QTLs drive experience? |

ior. Particularly useful are SNPs in coding regions (cSNPs) and other SNPs that are potentially functional, such as SNPs in DNA regions that regulate the transcription of genes. Methods are being developed to exploit genome-wide strategies to identify QTLs associated with complex traits (Craig and McClay 2003).

When the working draft of the human genome sequence was published, much publicity was given to the finding that there appear to be fewer than half as many genes in the human genome as expected – about 30,000 to 40,000 genes, similar to the estimates for mice and worms. A bizarre spin in the media was that having fewer genes somehow implies that nurture must be more important than we thought, because there are not enough genes to go around. There is, however, an important implication that follows from the finding that the human species does not have more genes than many other species: the number of genes is not responsible for the greater complexity of the human species. In part, the greater complexity of the human species occurs because, during the process of decoding genes into proteins, human genes more than the genes of other species are spliced in alternative ways to create a greater variety of proteins. Such subtle variations in genes, rather than the number of genes, may be responsible for differences between mice and men. If subtle DNA differences are responsible for the differences between mice and men, even more subtle differences are likely to be responsible for individual differences within the species.

Another relevant finding from the Human Genome Project is that less than 2% of the three billion letters in our DNA code involves genes in the traditional sense, that is, genes that code for amino acid sequences. This 2% figure is similar in other mammals that have been studied. On an evolutionary time scale, mutations are quickly weeded out from these bits of DNA by natural selection, because their code is so crucial for development. When mutations are not weeded out, they can cause one of the thousands of severe but rare single-gene disorders. However, it seems increasingly unlikely that the other 98% of DNA is just along for the ride. For example, some variations in this other 98% of the DNA are known to regulate the activity of the 2% of the DNA that codes for amino acid sequences. For this reason, the other 98% of DNA might be a good place to look for genes associated with quantitative rather than qualitative effects on behavioral traits.

## Behavioral Genomics

The future for behavioral genetics looks brighter than ever in the dawn of a postgenomic era. Behavioral genetics will be the major beneficiary of postgenomic developments that will facilitate investigation of complex traits influenced by many genes as well as by many environmental factors, first for finding genes associated with behavior and then for understanding the mechanisms by which those genes affect behavior at all levels of analysis, from the cell to the brain to the whole organism. Behavioral disorders and dimensions are the ultimate complex traits and will be swept along in the wake of the Human Genome Project as it increasingly provides the tools needed for the genetic analysis of complex traits.

Despite the slower-than-expected progress to date in finding genes associated with behavior, the substantial heritability of behavioral dimensions and disorders means that DNA polymorphisms exist that affect behavior. I am confident that more specific genes will be found for behavioral dimensions and disorders. Although attention is now focused on finding genes associated with complex traits, the greatest impact for behavioral science will come after genes have been identified. Few behavioral scientists are likely to join the hunt for genes because it is difficult and expensive, but once genes are found, it is relatively easy and inexpensive to use them. DNA can be obtained painlessly and inexpensively from cheek swabs; blood is not necessary. Cheek swabs yield enough DNA to genotype thousands of genes, and the cost of genotyping is surprisingly inexpensive when only a few DNA markers are genotyped. New techniques will make it increasingly inexpensive and easy to genotype large numbers of subjects for many thousands of genes.

Behavioral genetics is likely to accelerate in the post-genomic era as behavioral scientists incorporate DNA polymorphisms in their research, capitalizing on the new scientific horizons for understanding behavior. What has happened in the area of dementia in the elderly will be played out in many other areas of the behavioral sciences. The only known risk factor for late-onset Alzheimer's dementia (LOAD) is a gene, apolipoprotein E, involved in cholesterol transport. A form of the gene called allele 4 quadruples the risk for LOAD but is neither necessary nor sufficient to produce the disorder; hence, it is a QTL. Although the association between allele 4 and LOAD was reported less than a decade ago (Corder et al. 1993), it has already become de rigueur in research on dementia to genotype subjects for apolipoprotein E to ascertain whether the results differ for individuals with and without this genetic risk factor. Genotyping apolipoprotein E will become routine clinically if this genetic risk factor is found to predict differential response to interventions or preventative treatments.

In terms of clinical work, DNA might eventually contribute to gene-assisted diagnoses and treatment programs. The most exciting potential for DNA research is to use DNA as an early warning system that facilitates the development of primary interventions that prevent or ameliorate disorders before they create collateral damage. These interventions for behavioral disorders, and even for single-gene disorders, are likely to be behavioral rather than biological, involving environmental rather than genetic engineering. For example, phenylketonuria (PKU), a metabolic disorder that results postnatally in severe mental retardation, is caused by a single gene on chromosome 12. This form of mental retardation has been largely prevented, not by high-tech solutions such as correcting the mutant DNA or by eugenic programs or by drugs, but rather by a change in diet that prevents the mutant DNA from having its damaging effects. Because this environmental intervention is so cost-effective, newborns have been screened for PKU for decades to identify those with the disorder so that their diet can be changed. The example of PKU serves as an antidote to the mistaken notion that genetics implies therapeutic nihilism, even for a single-gene disorder. This point is even more relevant to complex disorders that are influenced by many genes and by many environmental factors as well. With behavior-based interventions, psychotherapists will eventually be in the business of preventing the consequences of gene expression.

The search for genes involved in behavior has led to a number of ethical concerns. For example, there are fears that the results will be used to justify social inequality, to select individuals for education or employment, or to enable parents to pick and choose among their fetuses. These concerns are largely based on misunderstandings about how genes affect complex traits (Rutter and Plomin 1997), but it is important that behavioral scientists knowledgeable about DNA continue to be involved in this debate. Students in the behavioral sciences must be taught about genetics to prepare them for this post-genomic future.

The future of genetic research will involve moving from finding genes (genomics) to finding out how genes work (functional genomics). Functional genomics is generally viewed as a bottom-up strategy, in which gene products are identified by their DNA sequences and the functions of the gene products are traced through cells and then cell systems and eventually the brain. This view is captured by the term "proteomics," which focuses on the function of gene products (proteins). But there are other levels of analysis at which we can understand how genes work. At the other end of the continuum is a top-down level of analysis that considers the behavior of the whole organism. For example, we can ask how the effects of specific genes unfold in behavioral development and how they interact and correlate with experience. This top-down, behavioral level of analysis is likely to pay off more quickly in prediction, diagnosis, and intervention than the slow build-up of knowledge through cell systems. The phrase "behavioral genomics" has been suggested to emphasize the potential contribution of top-down levels of analysis towards understanding how genes work (Plomin and Crabbe 2000). Bottom-up and top-down levels of analysis of gene-behavior pathways will eventually meet in the brain. The grandest implication for science is that DNA will serve as an integrating force across diverse disciplines.

# References

Baron M (2001) Genetics of schizophrenia and the new millennium: progress and pitfalls. Am J Human Genet 68: 299–312

Capecchi MR (1994) Targeted gene replacement. Sci Am 270 : 52–59

Cardon LR, Bell J (2001) Association study designs for complex diseases. Nat Genet 2: 91–99

Cardon LR, Smith SD, Fulker DW, Kimberling WJ, Pennington BF, DeFries JC (1994) Quantitative trait locus for reading disability on chromosome 6. Science 266: 276–279

Carroll JB (1993) Human cognitive abilities. Cambridge University Press, New York

Corder EH, Saunders AM, Strittmatter WJ, Schmechel DE, Gaskell PC, Small GW, Roses AD, Haines JL, Pericak Vance MA (1993) Gene dose of apolipoprotein E type 4 allele and the risk of Alzheimer's disease in late onset families. Science 261: 921–923

Craig I, McClay J (2003) The role of molecular genetics in a post genomics world. In: Plomin R, DeFries JC, Craig IW, McGuffin P (eds) Behavioral genetics in a postgenomic era. APA Books, Washington, DC, pp 19–40

Deater-Deckard K, Reiss D, Hetherington EM, Plomin R (1997) Dimensions and disorders of adolescent adjustment: A quantitative genetic analysis of unselected samples and selected extremes. J Child Psychol Psychiat 38: 515–525

DeFries JC, Gillis JJ (1993) Genetics and reading disability. In: Plomin R, McClearn GE (eds) Nature, nurture and psychology. American Psychological Association, Washington, DC, pp 121–145

Duncan J, Seitz RJ, Kolodny J, Bor D, Herzog H, Ahmed A, Newell FN, Emslie H (2000) A neural basis for general intelligence. Science 289: 457–460

Fisher RA (1918) The correlation between relatives on the supposition of Mendelian inheritance. Trans Roy Soc Edinburgh 52: 399–433

Gazzaniga MS (ed) (2000) Cognitive neuroscience: a reader. Blackwells, Oxford

Kendler KS, Eaves LJ (1986) Models for the joint effects of genotype and environment on liability to psychiatric illness. Am J Psychiat 143: 279–289

Lesch PK (2003) Neuroticism and serotonin: a developmental genetic perspective. In: Plomin R, DeFries JC, Craig IW, McGuffin P (eds) Behavioral genetics in a postgenomic era. APA Books, Washington, DC, pp 389–424

McGue M, Bouchard TJ, Jr., Iacono WG, Lykken DT (1993) Behavioral genetics of cognitive ability: A life-span perspective. In: Plomin R, McClearn GE (eds) Nature, nurture, and psychology. American Psychological Association, Washington, DC, pp 59–76

Plomin R (1994) Genetics and experience: The interplay between nature and nurture. Sage Publications Inc., Thousand Oaks, California

Plomin R (1999) Genetics and general cognitive ability. Nature 402: C25–C29

Plomin R, McClearn GE (1993) Nature, nurture, and psychology. American Psychological Association, Washington, DC

Plomin R, Crabbe JC (2000) DNA. Psych Bull 126: 806–828

Plomin R, Owen MJ, McGuffin P (1994) The genetic basis of complex human behaviors. Science 264: 1733–1739

Plomin R, Asbury K, Dunn J (2001a) Why are children in the same family so different? Nonshared environment a decade later. Can J Psychiat 46: 225–233

Plomin R, DeFries JC, McClearn GE, McGuffin P (2001b) Behavioral genetics. 4th Edition. Worth Publishers, New York

Plomin, R., Spinath, F. M. (2002) Genetics and general cognitive ability (g). Trends Cogn Sci, 6: 169–176

Risch NJ (2000) Searching for genetic determinants in the new millennium. Nature 405: 847–856

Risch N, Merikangas KR (1996) The future of genetic studies of complex human diseases. Science 273: 1516–1517

Rowe DC, Jacobson KC, Van den Oord JCG (1999) Genetic and environmental influences on vocabulary IQ: Parental education level as moderator. Child Dev 70: 1151–1162

Rutter M, Plomin R (1997) Opportunities for psychiatry from genetic findings. Brit J Psychiat 171: 209–219

Seligman MEP, Csikszentmihalyi M (2000) Positive psychology: An introduction. Am Psychol 55: 5–14

Thapar A (2003) Attention deficit hyperactivity disorder: new genetic findings, new directions. In: Plomin R, DeFries JC, Craig IW, McGuffin P (eds) Behavioral genetics in a postgenomic era. APA Books, Washington, DC, , pp 445–462

Thompson PM, Cannon TD, Narr KL, van Erp T, Poutanen VP, Huttunen M, Lonnqvist J, Standertskjold-Nordenstam CG, Kaprio J, Khaledy M, Dail R, Zoumalan CI, Toga AW (2002) Genetic influences on brain structure. Nature Neurosci, 4: 1253–1258

Thompson RF (2000) The brain: a neuroscience primer. Worth, New York

Wahlström J (1990) Gene map of mental retardation. J Mental Deficiency Research 34: 11–27

Wehner JM, Balogh SA (2003) Genetic studies of learning and memory in mouse models. In: Plomin R, DeFries JC, Craig I, McGuffin P (eds) Behavioral genetics in a postgenomic world. APA Books, Washington, DC, pp 103–122

Willcutt EG, DeFries JC, Pennington BF, Smith SD, Cardon LR, Olson RK (2003) Comorbid reading difficulties and ADHD. In: Plomin R, DeFries JC, Craig IC, MuGuffin P (eds) Behavioral genetics in a postgenomic era. APA Books, Washington, DC, 227–246

# The Genetic Basis of a Severe Speech and Language Disorder

S.E. Fisher[1]

## Summary

The KE family represent the only documented case of single-gene inheritance of a speech and language disorder. There has been some debate over the specificity of their impairment and the precise nature of the core deficit. Nevertheless, it is generally agreed that the gene that is disrupted in affected members of this family must play a key role in neurological mechanisms that are important for speech and language acquisition. The simple transmission pattern of the difficulties in the KE family allowed geneticists to use a traditional strategy to map the gene responsible to a small interval on chromosome 7. They then exploited data from large-scale human genomic sequencing efforts to assemble a detailed map of genes in this chromosomal region. A child was identified (unrelated to the KE family) who has speech and language disorder associated with a gross chromosomal abnormality involving the candidate region of chromosome 7. It was demonstrated that the abnormality in this child directly interrupts a novel gene encoding a polyglutamine repeat and a forkhead/winged-helix DNA-binding domain. The gene, known as FOXP2, is strongly expressed in the developing brain during embryogenesis and belongs to a large family of transcription factors involved in switching on and off other genes. Mutation screening of FOXP2 in the KE family revealed a point mutation in all affected individuals, which leads to alteration of a key residue in the DNA-binding domain, and is predicted to disrupt the function of the protein. In the future, studies of FOXP2 may provide a unique entry-point for investigating molecular processes mediating speech and language development.

## Introduction

In the last decade there has been considerable interest in investigations of an intriguing three-generational family from the United Kingdom known as KE. About half of the members of this family are affected with a severe disability that appears to be passed down from generation to generation in a simple fashion

---

[1] Wellcome Trust Centre for Human Genetics, Oxford University, Roosevelt Drive, Oxford, OX3 7BN, UK; e-mail: simon.fisher@well.ox.ac.uk

Mallet/Christen
Neurosciencess at the Postgenomic Era
© Springer-Verlag Berlin Heidelberg 2003

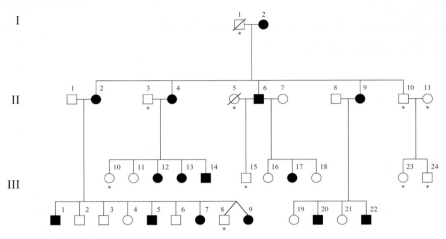

**Fig. 1.** Pedigree diagram of family KE. Squares represent males; circles represent females. A diagonal line indicates that the individual is deceased. Family members with speech and language impairment are shaded. Asterisks indicate those individuals who were unavailable for linkage analysis. From Fisher et al. (1998) © Nature America Inc. Reprinted with permission.

(Hurst et al. 1990). As can be seen from the pedigree diagram (Fig. 1), this is a classic textbook example of Mendelian inheritance with a dominant mode of transmission. In other words, the family history suggests that a mutation in a single gene accounts for the disorder, and that inheriting only one defective copy of this putative gene is enough to cause difficulties. On the face of it there is nothing odd about this observation; a great number of genetic disorders are known to have straightforward transmission patterns and, in the majority of cases, this fact has allowed geneticists to track down mutations that cause the disease (see Collins 1995). What makes this particular case so unusual is the nature of the disorder that is being inherited.

The affected individuals of the KE family have profound difficulties with acquiring speech and language and many of their problems persist into adulthood (Fisher et al. 1998). Such disorders are not unknown; up to 5% of children suffer from unexplained impairments in speech and language abilities, despite adequate intelligence and environmental stimulation (Bishop 2001). Furthermore, these kinds of problems do tend to run in families (Lewis et al. 1989; Tallal et al. 1989; Tomblin 1989), and studies of twins indicate that genetic factors are indeed likely to play an important role (Lewis and Thompson 1992; Bishop et al. 1995; Tomblin and Buckwalter 1998). However, the inheritance patterns of speech and language problems in affected families tend to be complex (Fisher 2002); there is no simple correspondence between the genetic makeup of an individual (genotype) and his or her cognitive abilities (phenotype). Some children with high-risk genotypes do not have any problems, whereas others who are at low risk genetically may nevertheless develop a disorder, for example, due to environmental factors. Furthermore there could be different genes contributing to speech and language problems in different families, and/or multiple genes interacting within a single family to increase risk. To date, the KE family represent the sole documented exception to this genetic complexity. Recently, Cecilia Lai, Simon Fisher and colleagues tracked down the specific mutation that is responsible for the disorder in this family (Lai et al. 2001). The implication that a simple genetic cause may underlie deficits in a unique human characteristic such as the ability to speak has generated a great deal of attention as well as a fair share of controversy (see Pinker 1994). This chapter will attempt to provide a balanced overview of the studies of the KE family and the implications of the recent findings.

## What is the nature of the disorder in the KE family?

Given the apparent monogenic transmission of the problems experienced by the KE family, it is not surprising that their disorder has been the subject of intense study by a number of research groups since they were first described in 1990. There has been a certain amount of debate revolving around issues like the specificity of language deficits and the general intelligence of the affected pedigree members (e.g., Gopnik and Crago 1991; Pinker 1994; Vargha-Khadem et al. 1995; Watkins et al. 2002), and readers of the literature may find the differing accounts somewhat confusing. The clear message from reviewing these studies is that, while their pathology is certainly restricted to the central nervous system, the KE family do not have a clean disruption of one highly specific feature of brain ability. Rather, they show deficits in several related aspects of cognition. In fact, this is exactly what we should expect; even classic single-gene disorders such as cystic fibrosis and Duchenne muscular dystrophy involve complex phenotypes with multiple features. For example, in cystic fibrosis, severe inflammatory lung disease is accompanied by salty sweat, pancreatic insufficiency, intestinal obstruction and male infertility, all caused by the same mutations in a single gene encoding a chloride transporting protein. Such mapping of simple inheritance to complex phenotype is commonly observed and there is no reason why this should not also be the case for the monogenic disorder of brain development observed in the KE family, especially given the particularly complex nature of the organ in question.

Some researchers have focused closely on the difficulties that affected KE individuals experience in aspects of grammar, such as their ability to manipulate words according to rules of tense, number and gender (Gopnik and Crago 1991). Results from such studies have been central to hypotheses regarding innate aspects of language acquisition (Pinker 1994). However, a more comprehensive description of the KE phenotype needs to acknowledge the following. Firstly, the disorder involves severe impairment in the selection and sequencing of the fine mouth movements that are necessary for speech, referred to as "verbal dyspraxia" (Hurst et al. 1990; Vargha-Khadem et al. 1995; Alcock et al. 2000). Note that this is a problem of orofacial motor *control* (i.e., despite normal facial musculature) and there is no evidence of any limb dyspraxia associated with the disorder (Watkins et al. 2002). Secondly, affected individuals have deficits in multiple aspects of language ability (such as the processing of units of speech) and grammatical skills (including use of word inflections and syntactical structure; Vargha-Khadem et al.,1995). These difficulties occur in receptive and expressive domains (although they are more profound for the latter) and are apparent for both spoken and written language (Watkins et al. 2002). Finally, it has been noted that the mean non-verbal intelligence quotient (IQ) of affected individuals is significantly lower than that of the unaffected members of the family (Vargha-Khadem et al. 1995; Watkins et al. 2002). However, non-verbal deficits do not segregate with the disorder in the pedigree; there are affected individuals who have non-verbal IQs that are in the normal range despite having severe speech and language difficulties (see Vargha-Khadem et al. 1995). Therefore, non-verbal cognitive deficits do not represent a central characteristic of the KE family disorder, although the relationship between verbal and non-verbal difficulties does merit further study (Watkins et al. 2002).

So, there remains some discussion over what might be considered the "primary" deficit in the particular form of disorder experienced by the KE family. Kate Watkins and colleagues (2002) suggest that the phenotypic profile may arise as a consequence of a basic impairment in the ability to sequence movement. Under this viewpoint, the development of linguistic and grammatical abilities are tightly intertwined with the ability to produce speech, such that articulation difficulties lead to impoverished language. However, such a hypothesis is challenged by studies of individuals with cerebral palsy, some of whom totally lack speech yet are able to demonstrate normal language and grammar skills if given an appropriate means of expression (Bishop et al. 1990).

## Genetic mapping of the SPCH1 gene to chromosome 7

Although researchers cannot currently agree on this issue of "core deficit," there is a general consensus that the gene disrupted in the KE family is likely to play a role in pathways that are *important* (even if not necessarily *specific*) for the development of speech and language. The observation of apparent monogenic inheritance in this pedigree meant that traditional methods of linkage analysis could be used to localise the risk gene to a small part of one chromosome. Linkage analysis involves examining variable markers in different chromosomal regions across the genome, for affected and unaffected members of a family. The inheritance patterns for such markers are compared to those seen for the disorder, and statistical tests are used to evaluate whether any markers "co-segregate" with the phenotypic trait. Such tests typically yield a LOD (logarithm of the odds ratio) score measuring the extent of linkage between a marker and a trait; for monogenic pedigrees a LOD score exceeding 3 (corresponding to 1000:1 odds in favour of linkage) is usually taken as significant evidence of marker-trait linkage. If there is significant marker-trait linkage in a specific chromosomal region, it is likely that the aetiological mutation will be found in a gene from this region, so this approach allows geneticists to use positional information to eventually pinpoint the mutation ("positional cloning;" see Collins 1995).

In most cases of language-related disorder the underlying genetic complexity is problematic for linkage analysis, although such difficulties are now beginning to be overcome with new statistical methods and high-throughput analyses of large numbers of families (e.g., Fisher et al. 1999a,b,; 2002; SLI consortium 2002). With the KE family there were no such difficulties, and Fisher and co-workers (1998) were thus able to use standard techniques to map the gene responsible for the disorder to a small interval of chromosome 7, in band q31. The linkage evidence was extremely convincing, generating a LOD score of 6.62, with 100% concordance between genotype and phenotype for several markers mapping in 7q31. The locus was assigned the name SPCH1 (i.e., speech disorder 1). This finding was the first formal proof that mutation of a single gene could lead to a speech and language disorder and represented an essential step towards its identification.

## Clues to aid the search for SPCH1

The linkage analyses described above were able to assign an approximate position for the SPCH1 locus on chromosome 7q31, but this region was likely to contain many genes, any of which might harbour the mutation causing the speech and language disorder. At this point in the study, there were two factors that proved to be crucial for future success.

1) At the time, the SPCH1 interval was being sequenced as part of the international effort to sequence the entire human genome. Using the available data, Lai et al. (2000) assembled a sequence-based map of the region and analysed it with bioinformatic software, including similarity search and gene prediction programs. This kind of *in silico* (i.e., at the computer) research is likely to form a major part of future genetic studies (Fisher 2002). The Lai et al. (2000) map contained nearly eight million base pairs of finished sequence, including 20 known genes and more than 50 anonymous transcripts of unknown function. The researchers began to sift through the genes, searching for mutations in the affected KE individuals. Those genes that might be predicted to encode proteins with a role in neurological development were prioritised for study. Of particular interest was a partially characterised gene encoding a long stretch of consecutive glutamine residues. Expansion of polyglutamine stretches has been implicated in a number of neurological disorders, such as Huntington chorea (Cummings and Zoghbi 2000), so this gene represented an attractive candidate for SPCH1.

2) Lai et al. (2000) identified a new case, unrelated to the KE family, where speech and language disorder is associated with a gross chromosomal abnormality involving the SPCH1 region. This patient, known as CS, has a translocation; part of the long arm of his chromosome 7 is exchanged with part of the long arm of his chromosome 5. The disorder of CS is notably similar to that observed in the KE family; he has severe verbal dyspraxia, substantial impairment of expressive and receptive language, and some lowering of IQ, which is more profound in the verbal domain. Lai et al. (2000) were able to demonstrate that the chromosome 7 breakpoint in CS lies in close proximity to the partially characterised polyglutamine repeat gene described above, adding further support to the idea that this might be the SPCH1 gene. However, analysis of all the sequence that was then known from this excellent candidate

did not reveal any mutations in affected members of the KE family (Lai et al. 2000).

## Finding the mutation in the KE family

Since the polyglutamine repeat gene represented such a promising candidate for SPCH1, but was only partially characterised, Lai et al. (2001) set about finding the remainder of the gene. To do this they exploited newly generated, large-scale genomic sequence data, which filled a gap in a crucial part of their sequence-based map of 7q31. Using these data, they were finally able to predict the complete structure of the polyglutamine gene and verify this using laboratory-based methods. The newly identified part of the gene encoded a characteristic DNA-binding motif, known as a forkhead/winged-helix domain (Fig. 2). This finding suggested that the gene was a novel member of the forkhead box (FOX) family of transcription factors, and it was therefore given the official name FOXP2 (Kaestner et al. 2000). There are a large number of different FOX proteins, many of which appear be critical regulators of gene expression during embryogenesis (Kaufmann and Knochel 1996). (The founding member of the FOX family is essential for proper formation of terminal structures in fruit-fly embryos, and the original "forkhead" name relates to the phenotype of flies which were mutant for that particular gene.) FOX genes have been found to be mutated or disrupted in a number of human disorders, such as glaucoma (FOXC1; Nishimura et al. 1998), thyroid agenesis (FOXE1; Clifton-Bligh et al. 1998), immune deficiency (FOXP3; Wildin et al. 2001; Bennett et al. 2001) and ovarian failure (FOXL2; Crisponi et al. 2001).

Lai et al. (2001) found that FOXP2 was expressed in a range of human foetal and adult tissues. Importantly, there was strong expression in the developing central nervous system of the foetus. Analysis of patient CS, who has the translocation involving chromosomes 5 and 7, confirmed that the chromosome-7 breakpoint directly disrupted the FOXP2 gene (Fig. 2) Lai et al. (2001) went on to sequence the newly identified regions of FOXP2 in the KE family, and discovered a point mutation in the DNA-binding domain that co-segregated perfectly with the speech and language disorder (i.e., all affected individuals had the mutation, all unaffected members of the family had the normal version of the gene). This change in the DNA sequence of the gene alters an arginine amino acid residue at a critical point of the FOXP2 protein, one that is invariant in all known FOX proteins, in diverse organisms such as yeast, fruit-fly, mouse and man. Furthermore, the change was not present in the chromosomes of a large number of unrelated humans. It is thus highly likely that this arginine residue is of key importance for FOXP2 function and that the mutation in the KE affected individuals is responsible for their disorder.

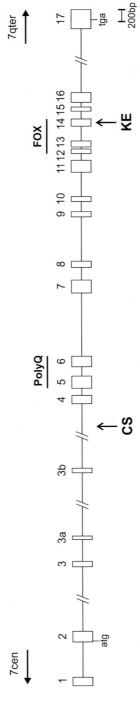

**Fig. 2.** Schematic representation of human FOXP2 gene structure. Boxes represent *exons* (present in the processed messenger RNA), lines represent *introns* (removed from the messenger RNA by splicing). Exons 3a and 3b are alternatively spliced (i.e., present in some forms of the messenger RNA, but not others). Positions of initiation (atg) and termination (tga) codons are indicated, which represent the start and end sites for protein translation, respectively. The scale shown applies only to exons; the entire region spans several hundred thousand bases of genomic DNA. The orientation with respect to the centromere (7cen) and long arm terminal end (7qter) of chromosome 7 is given. Exons encoding polyglutamine tracts (PolyQ) and the forkhead domain (FOX) are indicated. The CS translocation breakpoint lies between exons 3b and 4, directly disrupting the genomic locus. The point mutation in affected KE individuals maps within exon 14, altering the amino acid sequence of the FOX domain. From Lai et al. (2001) © Nature America Inc. Adapted with permission.

## Molecular dissection of neurological pathways

Lai et al. (2001) proposed that the speech and language disorder in the KE family, and in translocation case CS, results from insufficient quantities of functional FOXP2 protein at some specific stage of neural development. Although it must be recognised that this is only a working hypothesis, it suggests that FOXP2 may give us an exciting and unique entry-point into molecular mechanisms underlying speech and language development. As such, there are many interesting questions that immediately arise from these studies, some of which are already being addressed. For example, Newbury et al. (2002) have investigated whether mutations in FOXP2 may contribute to more general and genetically complex forms of speech and language impairment, but conclude that this gene is unlikely to be a common determinant of such problems. However, it will be possible in the coming years to identify the downstream targets of FOXP2: those genes which it switches on and off during neural development. Some of these target genes might be considered in the future as candidates for complex speech and language disorders. Additional studies are focussing on the expression patterns of FOXP2 in the developing nervous system. It will be of great interest to compare sites of expression to sites of pathology suggested by brain imaging studies of speech and language disorder patients, although this may not necessarily reveal a simple correspondence. Finally, researchers are using gene-targeting technology to explore the role of FOXP2 in mouse embryogenesis and evaluating the molecular evolution of the gene in primates to look for evidence of selection. In conclusion, it is worth noting that many genetic factors are likely to be involved in the neurological processes that mediate speech and language acquisition. Nevertheless, the discovery of FOXP2 represents the first piece of a complex puzzle, which will no doubt lead to many fascinating insights over the years to come.

## References

Alcock KJ, Passingham RE, Watkins KE, Vargha-Khadem F (2000) Oral dyspraxia in inherited speech and language impairment and acquired dysphasia. Brain Lang 75: 17–33

Bennett CL, Christie J, Ramsdell F, Brunkow ME, Ferguson PJ, Whitesell L, Kelly TE, Saulsbury FT, Chance PF, Ochs HD (2001) The immune dysregulation, polyendocrinopathy, enteropathy, X-linked syndrome (IPEX) is caused by mutations of FOXP3. Nature Genet 27: 20–21

Bishop DVM (2001) Genetic and environmental risks for specific language impairment in children. Phil Trans Biol Sci 356: 369–380

Bishop DVM, Brown BB, Robson J (1990) The relationship between phoneme discrimination, speech production, and language comprehension in cerebral-palsied individuals. J Speech Hear Res 33: 210–219

Bishop DVM, North T, Donlan C (1995) Genetic basis for specific language impairment: evidence from a twin study. Dev Med Child Neurol 37: 56–71

Clifton-Bligh RJ, Wentworth JM, Heinz P, Crisp MS, John R, Lazarus JH, Ludgate M, Chatterjee VK (1998) Mutation of the gene encoding human TTF-2 associated with thyroid agenesis, cleft palate and choanal atresia. Nature Genet 19: 399–401

Collins FS (1995) Positional cloning moves from perditional to traditional. Nature Genet 9: 347–350

Crisponi L, Deiana M, Loi A, Chiappe F, Uda M, Amati P, Bisceglia L, Zelante L, Nagaraja R, Porcu S, Ristaldi MS, Marzella R, Rocchi M, Nicolino M, Lienhardt-Roussie A, Nivelon A, Verloes A, Schlessinger D, Gasparini P, Bonneau D, Cao A, Pilia G (2001) The putative forkhead transcription factor FOXL2 is mutated in blepharophimosis/ptosis/epicanthus inversus syndrome. Nature Genet 27: 159–166

Cummings CJ, Zoghbi, HY (2000) Fourteen and counting: unraveling trinucleotide repeat diseases. Hum Mol Genet 9: 909–916

Fisher SE (2002) Isolation of the genetic factors underlying speech and language disorders. In: Plomin R, DeFries JC, Craig IW, McGuffin P (eds) Behavioral genetics in the postgenomic era. APA Books, Washington DC, USA, 205–226

Fisher SE, Vargha-Khadem F, Watkins KE, Monaco AP, Pembrey, ME (1998) Localisation of a gene implicated in a severe speech and language disorder. Nature Genet 18: 168–170

Fisher SE, Marlow AJ, Lamb J, Maestrini E, Williams DF, Richardson AJ, Weeks DE, Stein JF, Monaco AP (1999a) A quantitative-trait locus on chromosome 6p influences different aspects of developmental dyslexia. Am J Human Genet 64: 146–156

Fisher SE, Stein JF, Monaco AP (1999b) A genome-wide search strategy for identifying quantitative trait loci involved in reading and spelling disability (developmental dyslexia). Eur Child Adolesc Psych 8(S3): 47–51

Fisher SE, Francks C, Marlow AJ, MacPhie IL, Newbury DF, Cardon LR, Ishikawa-Brush Y, Richardson AJ, Talcott JB, Gayán J, Olson RK, Pennington BF, Smith SD, DeFries JC, Stein JF, Monaco AP (2002) Independent genome-wide scans identify a chromosome 18 quantitative-trait locus influencing dyslexia. Nature Genet 30: 86–91

Gopnik M, Crago MB (1991) Familial aggregation of a developmental language disorder. Cognition 39: 1–50

Hurst JA, Baraitser M, Auger E, Graham F, Norell S (1990) An extended family with a dominantly inherited speech disorder. Dev Med Child Neurol 32: 347–355

Kaestner KH, Knöchel W, Martinez DE (2000) Unified nomenclature for the winged helix/forkhead transcription factors. Genes Dev 14: 142–146

Kaufmann E, Knöchel W (1996) Five years on the wings of fork head. Mech Dev 57: 3–20

Lai CSL, Fisher SE, Hurst JA, Levy ER, Hodgson S, Fox M, Jeremiah S, Povey S, Jamison DC, Green ED, Vargha-Khadem F, Monaco AP (2000) The SPCH1 region on human 7q31: genomic characterization of the critical interval and localization of translocations associated with speech and language disorder. Am J Human Genet 67: 357–368

Lai CSL, Fisher SE, Hurst, JA, Vargha-Khadem F, Monaco AP (2001) A novel forkhead-domain gene is mutated in a severe speech and language disorder. Nature 413: 519–523

Lewis BA, Thompson LA (1992) A study of developmental speech and language disorders in twins. J Speech Hear Res 35: 1086–1094

Lewis BA, Ekelman BL, Aram DM (1989) A familial study of severe phonological disorders. J Speech Hear Res 32: 713–724

Newbury DF, Bonora E, Lamb JA, Fisher SE, Lai CS, Baird G, Jannoun L, Slonims V, Stott CM, Merricks MJ, Bolton PF, Bailey AJ, Monaco AP, International Molecular Genetic Study of Autism Consortium (2002) FOXP2 is not a major susceptibility gene for autism or specific language impairment. Am J Human Genet, 70: 1318–1327

Nishimura DY, Swiderski RE, Alward WL, Searby CC, Patil SR, Bennet SR, Kanis AB, Gastier JM, Stone EM, Sheffield VC (1998) The forkhead transcription factor gene FKHL7 is responsible for glaucoma phenotypes which map to 6p25. Nature Genet 19: 140–147

Pinker S (1994) The language Iistinct. Allen Lane, London

SLI Consortium (2002) A genomewide scan identifies two novel loci involved in Specific Language Impairment (SLI). Am J Human Genet 70: 384–398

Tallal P, Ross R, Curtiss S (1989) Familial aggregation in specific language impairment. J Speech Hear Disord 54: 167–173

Tomblin JB (1989) Familial concentration of developmental language impairment. J Speech Hear Disord 54: 287–295

Tomblin JB, Buckwalter PR (1998) Heritability of poor language achievement among twins. J Speech Lang Hear Res 41: 188–199

Vargha-Khadem F, Watkins K, Alcock K, Fletcher P, Passingham R (1995) Praxic and nonverbal cognitive deficits in a large family with a genetically transmitted speech and language disorder. Proc Natl Acad Sci USA 92: 930–933

Watkins KE, Dronkers NF, Vargha-Khadem F (2002) Behavioural analysis of an inherited speech and language disorder: comparison with acquired aphasia. Brain 125: 452–464

Wildin RS, Ramsdell F, Peake J, Faravelli F, Casanova JL, Buist N, Levy-Lahad E, Mazzella M, Goulet O, Perroni L, Bricarelli FD, Byrne G, McEuen M, Proll S, Appleby M, Brunkow ME (2001) X-linked neonatal diabetes mellitus, enteropathy and endocrinopathy syndrome is the human equivalent of mouse scurfy. Nature Genet 27: 18–20

# Genetics and Physiopathology
# of X-linked Mental Retardation

*J. Chelly*[1] *and J.L. Mandel*[2]

## Summary

Mutations in X-linked genes account for an excess of males affected with mental retardation. Target genes have recently been identified both for syndromic forms of X-linked mental retardation and in families affected with "nonspecific" forms, where cognitive impairment is the only clinical feature. The latter are genetically very heterogeneous, as the eight genes identified up to know account for only a minority of affected families. Proteins that have a role in chromatin remodelling are affected in three important syndromic forms, while defects in signal transduction pathways implicated in neuronal maturation were found in "nonspecific" forms. These findings provide important insights into the molecular and cellular defects that underlie mental retardation.

## Introduction

Mental retardation (MR) is the most frequent cause of serious handicap in children and young adults. MR is defined as an overall "intelligence quotient" (IQ) lower than 70 associated with functional deficits in adaptive behavior (such as daily-living skills, social skills and communication), with an onset before 18 years (see Moser et al. 1990 and Stevenson et al. 2000) for review. Moderate to severe MR (IQ<50) is estimated to affect 0.4–0.8% of the population and the prevalence increases to 2% if mild MR (50<IQ<70) is included, although these estimates vary widely between epidemiological studies (Stevenson et al. 2000; McLaren and Bryson 1987).

The underlying causes of MR are extremely heterogeneous. There are several nongenetic factors that act prenatally or during early infancy and cause brain injury: these include infectious diseases, perinatal anoxia and maternal intoxication during pregnancy (which gives rise to fetal alcohol syndrome). Well-established genetic causes of MR include visible chromosomal anomalies and monogenic diseases; a search for MR in the Online Mendelian Inheritance in Man

[1] Laboratoire de Génétique et Physiopathologie des Retards Mentaux, IC, CHU Cochin 75014 Paris, France
[2] Institut des Génétique et de Biologie Moléculaire et Cellulaire, CNRS/INSERM/Université Louis Pasteur, Illkirch, CU Strasbourg, France

Mallet/Christen
Neurosciencess at the Postgenomic Era
© Springer-Verlag Berlin Heidelberg 2003

(OMIM) database identifies close to 1000 entries. However, 25–40% of cases with severe MR, and the vast majority of mild MR cases, remain unexplained. A large proportion of cases probably involves combinations of multigenic and environmental factors.

Our knowledge of the monogenic causes of MR has increased dramatically in recent years, notably through the use of positional cloning strategies that allow the identification of many disease genes associated with MR. Impairment of cognitive functions can be observed in metabolic diseases (such as the Smith-Lemli Opitz syndrome, a defect in cholesterol metabolism, or untreated phenylketonuria), as well as in developmental defects or neuromuscular diseases (such as holoprosencephaly or Duchenne muscular dystrophy, two entities in which MR is not a constant feature). Chromosome microdeletion syndromes are another important genetic cause of MR (such as the Prader-Willi, Angelman and Smith-Magenis syndromes) or of more subtle cognitive impairment (Williams syndrome; Shaffer et al. 2001; Donnai and Karmiloff - Smith 2000). Very recent studies suggest that chromosomal rearrangements that affect the telomeric regions of autosomes and are not detectable by conventional cytogenetic analysis may account for up to 7% of moderate to severe MR (Knight et al. 1999). Some diseases for which the gene is identified affect relatively large numbers of patients and families, such as the Fragile X syndrome (which affects 1 in 4000 males and 1 in 7000 females) and Rett syndrome (1 in 12000 girls), but our knowledge of monogenic causes of MR is still far from complete. Identification of the genes involved, especially in those cases where MR is the only obvious symptom, is one of the major challenges for human genetics for the years ahead. It is anticipated that this group of disorders is caused by alterations in molecular pathways that are important for cognitive functions, and that defining the genes involved will improve our understanding of te biological and cellular mechanisms underlying cognition.

In recent years, striking progress has been made in the identification of some of the X-linked genes associated with MR. This is because of the relative ease of studying X-linked traits. The identification of these genes has provided important insights into molecular and cellular dysfunctions that underlie MR. We review this progress, focusing on the implication of chromatin remodelling and signal transduction pathways, and discuss the work that remains to be done. In addition, we consider the present difficulties in applying some of these discoveries to diagnosis and genetic counselling in affected families.

## X-linked mental retardation

Since 1890, many studies have reported an excess of males in institutions or special schools caring for the mentally handicapped (reviewed in Stevenson et al. 2000). This excess was first thought to be due to societal biases, based on the assumption that affected girls might be preferentially kept at home. However, the observation of large families with a clear X-linked inheritance of MR, and improved epidemiological studies, led to the gradual acceptance that this 20–30% male excess might be due to mutations in X-linked genes. The identification of the Fragile X syndrome as a distinct clinical entity in the late 1970s was an important step in this process. This syndrome is associated with a specific clinical phenotype and accounts for ~2–3% of MR in males and for ~1% in females (who are on average less affected than males; Hagerman and Cronister 1996; Imbert et al. 1998). The interest in Fragile X syndrome led to the description of further X-linked MR (XLMR) syndromes and of several large families where MR is not associated with a specific clinical or metabolic phenotype ("nonspecific" or "nonsyndromic" XLMR; Stevenson et al. 2000). Syndromic forms of XLMR are, in general, amenable to conventional positional cloning strategies, because families sharing similar clinical phenotypes can be pooled for linkage analysis. However, the situation is much more complex for nonspecific XLMR. The first linkage studies demonstrated extensive heterogeneity; the MR genes (designated MRX genes) in different families mapped to different locations along the X chromosome. Linkage analysis of over 70 families showed that at least 10 nonoverlapping regions containing an MRX gene can be defined (Chiurazzi et al. 2001). The candidate region for each familiy is very large, usually in the 20–30 cM range, and might contain 100–300 genes (Fig. 1). Therefore, one cannot pool linkage results, even from families in which MRX genes map to overlapping regions, because these families might carry mutations in different genes.

Althought the nonsyndromic forms of XLMR do present a greater challenge to geneticists, progress towards identification of the underlying genetic causes has been made for these disorders as well as for syndromic XLMR. We first review some recent progress in the understanding of important forms of syndromic XLMR.

## Chromatin remodelling and XLMR

Genes have been identified for at least 20 XLMR diseases with distinctive symptomatology and varying degrees of MR (Table 1). Perhaps the most famous of these is the Fragile X syndrome, which is the most common cause of inherited MR and is associated with a chromosomal fragile site in Xq27.3. The causative mutation is an expansion of a CGG trinucleotide repeat that disrupts the expression of the FMR1 gene (Hagerman and Cronister 1996; Imbert et al. 1998). The FMR1 protein is an RNA-binding protein that has been proposed to be involved in regulating the transport or translation of specific mRNAs (for review see Bardoni et al. 2000). Some of the other disorders can be classified as metabolic diseases (Lesch-Nyhan syndrome, adrenoleukodystrophy, Hunter syndrome, Menkes disease), or as diseases associated with observable brain anomalies (X-linked hydrocephalus, X-linked lissencephaly, Pelizaeus-Merzbacher syndrome), and these are generally associated with prominent neurological manifestations. A more recent addition to the cellular processes that are disrupted in syndromic XLMR is chromatin remodelling. In this section, we review three disorders (ATRX, Coffin-Lowry and Rett syndromes) associated with severe MR, in which this process is thought to be affected by mutations in three distinct genes. Furthermore, although mutations in these genes were initially found in patients with the characteristic phenotype of each syndrome, mutations were subsequently

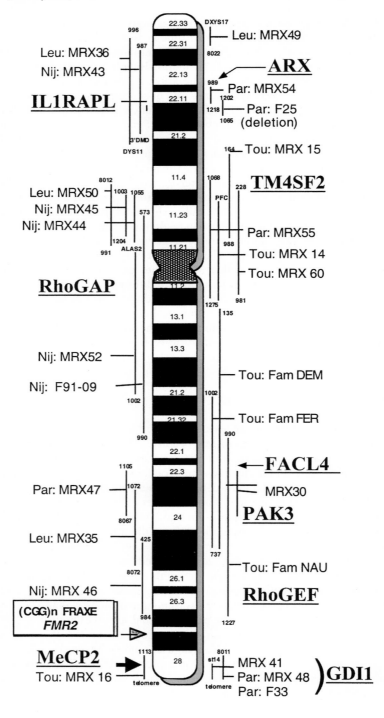

detected in some patients with non-specific MR. There are also interesting differences between the syndromes in the way they affect either sex: the ATRX syndrome affects exclusively males, Coffin-Lowry syndrome shows partial penetrance in females, and mutations in the MECP2 gene were initially detected in Rett syndrome, a disease that affects only girls.

## The ATRX syndrome

The ATRX gene was named because of its association with a form of very severe MR associated with biological signs of alpha thalassemia (HbH inclusions in red cells), in addition to characteristic facial features and frequent genital anomalies (Gibbons et al. 1995; Gibbons and Higgs 2000). The gene, identified in 1995, encodes a large nuclear protein that contains an N-terminal zinc finger, a coiled-coil domain and seven helicase motifs also found in helicases of the SNF2/SWI2 protein family, which are involved in chromatin remodelling (Gibbons et al. 1995;1997). Most of the mutations in this gene are missense truncating mutations (nonsense or frameshifts) are very rare and are limited to either end of the coding sequence. This suggests that most truncating mutations lead to prenatal lethality. The identification of ATRX mutations in more rare clinical syndromes (Juberg-Marsidi, Carpenter-Waziri, Holmes-Gang, and Smith-Fineman-Myers syndromes) provides an interesting example pf "syndrome lumping" (Villard et al. 1996, 2000; Stevenson 2000). While MR associated with mutations in ATRX is usually very severe (speech is absent or severely limited), a family was recently described in which two patients had the typical MR and facial features whereas two others had mild MR with epilepsy and no facial dysmorphism (Guerrini et al. 2000). Curiously, the mutation is a very N-terminal nonsense mutation, which indicates that the use of downstream ATGs for translation initiation might allow partial function. Females who carry an ATRX mutation show no symptoms, and this can be explained by selection for cells having the normal X active. Indeed, the mutated X chromosome was found to be inactivated in all leucocytes of carrier females (Gibbons et al. 1995; Lossi et al. 1999).

---

**Fig. 1.** Genes involved in nonspecific MR and genetic intervals corresponding to MRX families. Vertical bars indicate the candidate region for the MR gene mutated in each of 25 families (des Portes et al. 1999). The families are those studied in the European XLMR Consortium, with the exception of families MRX41 and MRX30 in which the first GDI1 and PAK3 mutations were found. Families in which a mutation was detected are marked by an asterisk (or in bold), The European XLMR Consortium was created in 1996, and involves the groups of J. Chelly (Paris), J. P Fryns (Leuven, Belgium), C. Moraine (Tours, France), H van Bockhoven and B.C.J. Hamel (Nijmegen, Netherlands) and H.H Ropers (Berlin, Germany). The collaborative efforts of the consortium have resulted in a collection of approximately 200 clinically well-characterized families with established or probable X-linked MR. Significant linkage to a subchromosomal region of the X chromosome (lod score >2) was obtained in more than 40 families (the most informative ones for such analysis). Lymphoblastoid cell lines were established for most probands of each family. DNAs and cell lines are exchanged between groups of the Consortium to test candidate genes

**Table 1.** Syndromic forms of X-linked mental retardation

| Disease[a] | Gene | Protein function |
|---|---|---|
| Adrenoleukodystrophy[b] | ALD | Peroxisomal ABC transporter involved in very long chain fatty acids catabolism |
| Coffin Lowry syndrome | RPS6KA3 | RSK2 serine-threonine protein kinase (RAS-MAP kinase signalling pathway) |
| Fragile X syndrome | FMR1 | mRNA binding protein, (control of mRNA transport/translation?) |
| Hunter disease (Mucopolysaccharidosis type II)[c] | IDS | Iduronate sulfatase (mucopolysaccharide catabolism) |
| X linked hydrocephalus/MASA syndrome | L1CAM | Adhesion molecule involved in neural cell interactions |
| Incontinentia Pigmenti (females)** | IKBKG | NEMO: NF kappa B essential modulator (LKK subunit-γ) |
| Lesch-Nyhan disease | HPRT | Hypoxanthine phosphoribosyl transferase (purine metabolism) |
| X-linked lissencephaly (males); subcortical laminar heterotopia (females) | DCX | Microtubule associated protein |
| Lowe oculo cerebrorenal syndrome | OCRL1 | Phosphoinositide phosphatase |
| Mental retardation and α thalassemia (X linked) | ATRX/XNP | DNA binding helicase, chromatin remodelling |
| Mohr-Tranebjaerg syndrome/DFN1 (deafness,blindness, dementia | DDP | Mitochondrial protein (protein targeting) import of proteins to mitochondrial new membrane |
| Norrie Disease** | NDP | Norrin secreted protein, growth factor? |
| Opitz G/BBB | MID1 | Ring box/B box protein associated to microtubules, regulated by protein phosphatase 2A |
| Rett syndrome (females) | MECP2 | Methyl CpG DNA binding protein, transcriptional repressor |
| West Syndrome (ISSX, MIM 308350) | ARX | Aristales-related homeobox protein |
| X-linked lissencephaly with abnormal genitalia (XLAG) | ARX | (Transcription factor) |

[a] Only those syndromes for which the gene has been identified are listed. For a list of other syndromes, see Stevenson et al. (2000).Syndromes in which MR is inconstant, and generally mild when present:Aarskog-Scott (FGDY), Dyskeratosis congenita (DKC1), Simpson Golabi-Behmel (GPC3), Duchenne Muscular Dystrophy (DMD), Pelizaeus-Merzbacher (PLP), Epilepsy with periventricular heterotopia (FLN1), PGK1 deficiency. The Menkes disease (copper transport), OTC (ornithine transcarbamylase) deficiency, and Pyruvate dehydrogenase deficiency are in general early lethal diseases. In these three diseases, survivor males, patients with milder forms or some heterozygous female carriers show impaired cognitive function.
[b] In ALD, intelligence is normal until the onset of brain demyelination, which occurs during childhood in only 40% of mutation-carrying males.
[c] In these diseases, patients with milder forms do not present cognitive impairment.

The ATRX protein is associated with pericentromeric heterochromatin and with the short arm of acrocentric chromosomes that contain ribosomal DNA arrays (McDowell et al. 1999). It interacts and colocalises with the heterochromatin protein HP1 and was reported to interact also with another protein implicated in chromatin remodelling, the homologue of the Drosophila enhancer of Zeste (Le Douarin et al. 1996; Cardoso et al. 1998; Berube et al. 2000). Finally, anomalies in the methylation of rDNA clusters and of a Y chromosome specific repeat were observed in ATRX patients (Gibbons et al. 2000). Although no obvious methylation differences were observed at the α globin locus, it should be noted that this locus is very close to the telomere, and its expression might be influenced by the abnormal control of heterochromatin in ATRX patients. Overall, the phenotype in ATRX syndrome seems likely to result from abnormalities in chromatin structure at certain regions, which in turn disrupts the expression of specific genes.

## Coffin-Lowry syndrome

Coffin-Lowry syndrome comprises severe MR, facial dysmorphism (which in some cases resembles that observed in patients with ATRX), and progressive skeletal malformations. Other, less frequent manifestations include deafness or cardiac defects.

The gene for Coffin-Lowry syndrome (RSK2/RPS6KA3) encodes a member of the RSK (for ribosomal protein S6 serine threonine kinase) family of protein kinases (Trivier et al. 1996). RSK2 is one of four RSK members in humans. The recently identified RSK4 gene is also X-linked and is a good candidate for an MR gene based on the analysis of Xq21 deletions in patients with contiguous gene syndromes that include deafness. This hypothesis has yet to be confirmed, as no point mutation in this gene has yet been found in MRX families (Yntema et al. 1999).

RSK2 protein comprises two ATP-binding kinase domains, regulatory phosphorylation sites and a docking site for the ERK kinase (a MAP kinase). Currently, 86 mutations have been identified in 250 patients with a clinical phenotype suggestive of Coffin-Lowry syndrome (Delaunoy et al. 2001). This relatively low hit rate is probably due to the rather nonspecific nature of some of the clinical symptoms, Mutations are very heterogeneous and include truncating mutations (60% of cases) and missense mutations (38% of cases). The mutational analysis has uncovered a larger spectrum of severity than initially expected, which prompted a search for RSK2 mutations in three MRX families mapping to the appropriate Xp22 region. In one family, a missense change was found very close to a regulatory phosphorylation site. This mutation leads to an 80% decrease in enzymatic activity of RSK2, and to a phenotype of mild MR, in the absence of other clinical features (Merienne et al. 1999). Thus, the residual RSK2 activity is thought to be sufficient to avoid severe MR and the skeletal manifestation of the disease. However, this is likely to be an exceptional cause of nonspecific XLMR. By contrast with ATRX syndrome, Coffin-Lowry mutations do not lead to biased inactivation and can therefore lead to clinical manifestations in carrier females. Indeed, 10% of RSK2 mutations have been identified in female probands with no affected male relatives, ascertained through learning disabilities and mild but suggestive facial and digital dysmorphism (Delaunoy et al. 2001).

The identification of the RSK2 gene as the gene for Coffin-Lowry syndrome was also useful for more fundamental biological studies. The very high sequence similarity between the various human RSK proteins made it very difficult to assess their functional specificity, in terms of upstream activators or downstream target genes. Analysis of fibroblasts from Coffin-Lowry patients showed that RSK2 was necessary for the activation by phosphorylation of the transcription factor CREB and for the induction of FOS expression, in response to epidermal growth factor (EGF) stimulation (De Cesare et al. 1998). RSK2 was also necessary for EGF- induced phosphorylation of histone H3, an important event for chromatin remodelling (Sassone-Corsi et al. 1999). Very recently, the interaction between RSK2 and the CBP coactivator (CREB binding protein, mutated in the Rubinstein-Taybi syndrome) was shown to regulate histone H3 acetylation (Merienne et al. submitted for publication). Thus, RSK2 appears to play an important role in chromatin remodelling events and gene regulation.

## From Rett syndrome to nonspecific MR

For many years, Rett syndrome, a severe disease that leads to cessation and regression of psychomotor development in girls, with accompanying autistic features, was an enigma in human genetics. The most popular hypothesis was that Rett syndrome is a dominant X-linked trait that causes prenatal lethality in boys. However, genetic analysis has been very difficult because almost all cases are sporadic. Nevertheless, very rare families with two or more affected girls allowed tentative mapping of the gene in Xq28 (Sirianni et al. 1998) and painstaking analysis of many genes in this region led to the identification of mutations in a gene that had been known for many years, and that encodes a methyl CpG DNA binding protein (MECP2; Amir et al. 1999). Mutations are found in 70 to 80% of girls who present the characteristic features of the disease (Amir and Zoghbi 2000). In the two years since initial identification of the Rett syndrome gene, more MECP2 mutations have been reported in more than 360 patients (see for instance Bourdon et al. 2001 and Laccone et al. 2001). The relatively high frequency of the disease appears linked to the presence of several mutation hot spots; seven CpG containing codons are implicated in 63% of the mutations, although there is no explanation for the very high mutability of these seven codons.

Furthermore, the C-terminal coding region is a target of small deletions found in 10% of patients. Mutations causing the typical Rett syndrome phenotype in girls lead to a complete loss of function of the MECP2 protein which will affect on average 50% of the cells of the patient (due to random X chromosome inactivation).

Once the Rett gene was identified, it was soon noticed that, in some families with an affected girl, boys were born with a very severe, and fatal, encephalopathy. Mutations in the MECP2 gene were found in such cases and this finding led to the view that the clinical spectrum due to MECP2 mutations, as for other XLMR genes, is broader than the initial definition of Rett syndrome. Indeed, MECP2 mutations were found in families with nonspecific male MR (Meloni et al. 2000; Orrico et al. 2000), and a recent study has shown that MECP2 mutations may account for less than 1% of males with MR (Couvert et al. 2001). Mutations found in these MR males are different from those observed in females with Rett syndrome and probably correspond to partial loss of MECP2 function,

MECP2 therefore represents the third XLMR gene to be implicated in chromatin remodelling (reviewed in Robertson and Wolffe 2000). The gene is ubiquitously expressed and so the particular sensitivity of neurons to its dysfunction remains a mystery. A postnatal neurologic phenotype reminiscent of Rett syndrome was recently observed in mouse MECP2 knock-out models (Chen et al. 2001; Guy et al. 2001), indicating that stability of brain function, rather than brain development, is sensitive to the absence of MECP2.

## Nonspecific MRX genes

For a long time, gene identification in nonspecific X-linked MR appeared to be an almost impossible task, given the extensive genetic heterogeneity of this condition. However, progress in genome analysis and the setting of efficient large collaborations between clinical and molecular teams (notably the European XLMR Consortium des Portes et al. 1999; see legend to Fig. 1) have led to spectacular progress in the past few years and the identification of seven MRX genes (eight if one includes MECP2; Table 2). But there might be as many as 30 still to find (Chiurazzi et al. 2001).

Two types of patients who carry X chromosomal rearrangements have been particularly useful in these studies: females with an X-autosome translocation, in whom the normal X chromosome is inactivated (a well-known feature of such translocations), with the result that all cells have a deficiency of an X-linked gene that lies at the translocation breakpoint; and males with microdeletions, often detected by the presence of a contiguous gene syndrome in which a known X-linked disease phenotype is combined with MR (in this case, one expects that an MR gene lies in the deleted region). In both cases, definitive proof that a gene is indeed an MR gene requires the identification of point mutations or intragenic small deletions in MRX families, or at least in sporadic cases of MR in males. Candidate gene strategies have also been fruitful, in which searches for mutations in MRX families have focused on genes located in the appropriate region of the X and that are known to be involved in neuronal development and function, or that are mutated in syndromic XLMR (such as MECP2 or RSK2).

### FMR2

Some mentally retarded patients present with a fragile site in Xq28 but not with the CGG expansion in the FMR1 gene that causes the fragile X syndrome. Such

**Table 2.** Genes involved in nonspecific X-linked mental retardation[a]

| Gene/location | Cloning strategy | Identified mutations[b] | Frequency of mutations in MR (%) | Potential function |
|---|---|---|---|---|
| FMR2/FRAXE Xq28 | Fragile site and deletions | CGG expansion rare deletions | ~0.2 | Transcription factor? |
| OPHN1 Xq12 | Breakpoint cloning | 1 PM | <0.5 | RhoGAP (RhoGTPase activating protein) regulation of actin cytoskeleton dynamics/neuronal morphogenesis |
| PAK3 Xq22 | Candidate gene | 2PM | <0.5 | p21 activating kinase 3 (Rac/Cdc42 effector) regulation of actin cytoskeleton dynamics/neuronal morphogenesis |
| GDI1 Xq28 | Candidate gene | 3 PM | <1 | RabGDP-dissociation inhibitor synaptic vesicle and activity, neuronal morphogenesis? |
| IL1RAPL1 Xp21.3-22.1 | Deletion mapping | Rare deletions 1 PM | <0.5 | IL-1 Receptor accessory protein-like, unknown function, synaptic plasticity (?) |
| TM4SF2 Xp11.4 | Breakpoint cloning | 2 PM | <0.5 | Member of the tetraspanin family, interacts with integrins, regulation of actin cytoskeleton dynamics (?) |
| ARHGEF6 Xq26 | Breakpoint cloning | 1 PM | <1 | Homologous to RhoGEF, effector of RhoGTPases regulation of actin cytoskeleton dynamics/neuronal morphogenesis |
| MECP2 Xq28 | Candidate gene (in male MR) | 6 PM (missenses) | 1–2 | Methyl CpG binding protein (see Table 1) |
| FACL4 Xq22.3 | deletion studies | 2 PM | <0.5 | Fatty acid-CoA ligase 4 |
| ARX XP22.1 | Candidate gene | missente mutation poly-alanine expansion | <0.5 | Aristaless related homeodomain protein |

[a] In addition, the RPS 6KA3/RSK2 and ATRX genes were found to be involved in nonspecific MR in single families, whereas mutations in these genes are otherwise associated with syndromic immer

[b] In independent families (PM, point mutations)

patients led to the identification of the FMR2 gene which is the target of a CCG repeat expansion and is associated with the FRAXE fragile site (Gu et al. 1996; Gecz et al. 1996). Deletion of the FMR2 gene was also noted in one patient with developmental and speech delay. The expansion mutation in FMR2 is located 600 kb downstream to the FMR1 gene and shares similar properties with the Fragile X/FMR1 mutation: the abnormal methylation of large repeats includes a CpG island and leads to extinction of transcription of FMR2, while intermediate sized expansions are unmethylated on the active X chromosome but are unstable, and are thus very similar to Fragile X premutations (Knight et al. 1994). Male patients with the methylated expansion usually present with mild/borderline nonspecific MR, although cases have been described who are either more severely affected or within the normal IQ range. The incidence of FRAXE/FMR2 expansion is about 10 times lower than that of the classic fragile X/FMR1 expansion (Yonings et al. 2000). FMR2 encodes a nuclear protein of unknown function, expressed in neurons, that belongs to a small group of proteins with DNA binding activity and that might function as transcription factors (Gecz et al. 1997; Miller et al. 2000).

## *RabGDI1*

The Rab GTPases are a subgroup (comprising at least 40 members) of the small Ras-like GTPase family and are involved in neurotransmission (Zerial and McBride 2001). In common with most small GTPases, Rab proteins cycle between an active, GTP-bound and an inactive, GDP-bound state through the action of regulatory proteins. GDP dissociation inhibitors (GDIs) are required to retrieve from the membrane the GDP-bound form and to maintain a pool of soluble Rab-GDP. In the mammalian brain, F128MaGDI encoded by the GDI1 gene is the most abundant form and regulates RAB3A and RAB3C, the Rab proteins that participate in synaptic vesicule fusion (geppert and Sudhof 1998). The mapping of GDI1 to Xq28 made the gene an excellent candidate for MRX families showing linkage to this region, and mutations were found in three of seven such families (D`Adamo et al. 1998; Bienvenu et al. 1998). One is a nonsense mutation, while the two others are missense mutations that decrease the affinity between RAB3A and GDI.

RabGDIα-deficient mice revealed a role for this protein in neurotransmitter release (Ishizaki et al. 2000). Furthermore, the phenotype of RabGDIα-deficient mice appears opposite to that of Rab3A-deficient mice; RabGDI α-deficiency leads to a sharp increase in facilitation of excitatory transmission during repetitive stimulation, whereas Rab3A deficiency leads to a decrease under the same conditions (Geppert et al. 1994). Recent data reported by Ishizaki et al. (2000) suggest that RabGDIα has an important role *in vivo* to suppress hyperexcitability of the pyramidal neurons.

## Three genes in the Rho GTPase pathway

Three of the newly identified MRX genes, OPHN1, PAK3 and ARHGEF, encode proteins that interact with Rho GTPases, a family of small Ras-like GTPases that act in signal transduction pathways from extracellular stimuli to the actin cytoskeleton and the nucleus (Fig. 2).

OPHN1 was found interrupted in a female patient carrying a X;12 balanced translocation associated with MR. A frameshift mutation causing premature termination was then identified in one of four MRX families showing linkage to the

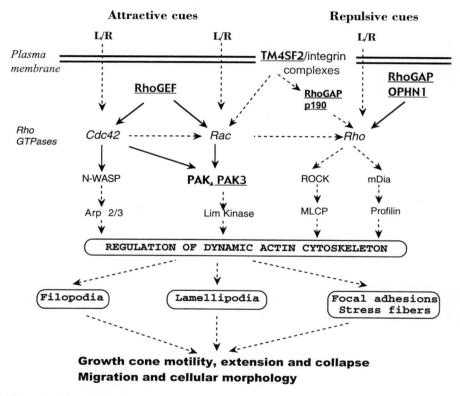

**Fig. 2.** Putative signal transduction pathways in growth cones and schematic illustration of potential interconnections between MRX genes through their involvement in the regulation of Rho family GTPases activity and actin cytoskeleton organization. Extracellular guidance cues (L) interact with growth cone receptors (R), which in turn activate signalling cascades involving Rho-like GTPases. Activated, GTP-bound, RhoGTpases stimulate filopodia and lamellipodia formation or induce growth cone collapse, Dysfunction of PAK3, RhoGAP, TM4SF2 and Rho-GEF proteins (bold and underlined proteins) is associated with MR. N-WASP, neuronal Wiskott-Aldrich syndrome protein; Arp2/3, actin-related protein 2/3; PAK; p-21 activating kinase. The tetraspanin/integrin complexes can activate Rac and Rho GTPases through mechanisms reviewed by Gioncotti (1997). Dotted arrows indicate oversimplified genetics.

appropriate X chromosome region (Billmart et al. 1998). OPHN1 encodes a protein (oligophrenin) that is similar to Rho GTPase activating proteins (RhoGAP) and stimulates GTPase activity only for members of the Rho family proteins, such as RhoA, Rac and Cdc42. These proteins are known to play a role in organisation of the cytoskeleton, and particulary in growth cone dynamics (Billuart et al. 1998). The transcript and the protein are mainly expressed in fetal and adult brain, in both neurons and glial cells. RhoGAP proteins increase the rate of GTP hydrolysis bound to Rho GTPases, and so loss of function of OPHN1 may result in constitutively active Rho proteins and alteration of actin cytoskeleton dynamics.

PAK3 is a member of the large family of p21 activating kinases (PAKs) and is highly expressed in developing brain. It was shown to act as a Rac/Cdc42 downstream effector (Bagrodia et al. 1995; Manser et al. 1995). Mutation screening of this candidate gene showed the presence of a nonsense mutation in one MRX family (Allen et al. 1998) and a missense mutation cosegregating with MR in another large family (Bienvenu et al. 2000). PAK proteins have been ascribed roles both in actin cytoskeleton dynamic regulation and in the Rac/Cdc42-induced activation of the Map kinase cascades.

The ARHGEF6 gene encodes a protein (also known as αPIX or Cool-2) with homology to guanine nucleotide exchange for RhoGTPases (Rho GEFs). Molecular analysis of a reciprocal X; 21 translocation in a male with MR showed that this gene was disrupted by the rearrangement (Kutsche et al. 2000). Mutation screening of 119 unrelated patients revealed a single intronic mutation in all the affected males in a large MRX family. The mutation causes preferential exon skipping and deletion of 28 amino acids (Kutsche et al. 2000). The role of ARHGEF6 in brain development and neuronal morphogenesis remains to be addressed, but it is required for PAK recruitment to actin cytoskeleton-rich structures such as focal complexes and lamellipodia, the formation of which is controlled by Cdc42 and Rac1 activity (Manser et al. 1995).

### TM4SF2

The TM4SF2 gene that encodes a tetraspanin (also known as TALLA-1/T cell acute lymphoblastic leukaemia antigen;Takagi et al. 19995) is inactivated by the Xp11.4 breakpoint of an X; 2 balanced translocation in a female patient with MR, and additionally point mutations were detected in two of 33 MR families (Zemni et al. 2000). Tetraspanins are cell-surface proteins of 200–300 amino acids that span the membrane four times and form two extracellular loops. One of the key features of the tetraspanins is their ability to associate with one another, with β1integrins and with class 1 and II HLA proteins. Their interaction with 1-integrins was suggested to mediate diverse cellular processes such as regulation of actin cytoskeleton dynamics, activation of signalling pathways, proliferation, adhesion and migration (Maecker et al. 1997). Very little is known about the role of tetraspanins in the physiology of CNS, where TM4SF2 appears highly expressed, notably in the cerebral cortex and hippocampus (Zemni et al. 2000).

## IL1RAPL

IL1RAPL (IL-1 receptor accessory protein like) was identified through investigation of a 350 kb deletion at Xp22.1-21.3 in an MRX familiy. The deletion overlapped with independent deletions in MR patients with contiguous gene syndromes that included glycerol kinase (GK) deficiency and adrenal hypoplasia (Carrie et al. 1999). Nonoverlapping deletions and a nonsense mutation in this large gene were identified in patients with cognitive impairment alone. The homologous mouse gene is expressed in the developing and postnatal structures of the hippocampus, which is implicated in learning and memory (Carrié et al. 1999). The ligand (s) that induces the cascade in which IL1RAPL is involved is still unknown, but preliminary data suggest a role in the regulation of synaptic vesicle exocytosis (Chelly et al. unpublished).

### FACL4: Fatty acid-CoA ligase 4

Contiguous gene deletion syndrome ATS-MR characterized by Alport syndrome (ATS) and mental retardation (MR) indicated Xq22.3 as a region containing one mental retardation gene (Meloni et al. 2002). Further investigation of the critical region for MR allowed the identification of two point mutations, one missense and one splice site change in the gene FACL4 in two families with nonspecific MR (Melonie et al. 2002). All carrier females with either point mutations or genomic deletions in *FACL4* showed a completely skewed X-inactivation, suggesting that the gene influences survival advantage. Acyl-CoA synthetases are a family of enzymes that catalyse the formation of acyl-CoA esters from fatty acids, ATP and coenzyme A. Five forms of fatty acid-CoA ligase have been identified in humans. *FACL4* encodes a protein of 670 amino acids expressed in several tissues, exept liver, the principal tissue of action of both *FACL1* and *FACL2*. In the brain, FACL4 encodes a longer transcript, resulting from alternative splicing, that produces a brain specific isoform containing 41 additional amino-terminal, hydrophobic amino acids. Data reported by Meloni et al. (Meloni et al. 2002) suggest that FACL4 protein is specifically expressed in neurons and not in glial cells. Although, it is difficult to speculate how the reduced production of arachidonyl-CoA esters could lead to mental retardation, involvement of these molecules in crucial processes such as regulation of $Ca^{2+}$ ions fluxes could (Knudsen et al. 1999) provide a basis for further investigation to understand mechanisms underlying MR.

### ARX: aristaless related homeobox gene

Investigation of a critical region for an X-linked mental retardation (XLMR) locus led to the identification of a novel Aristaless related homeobox gene (*ARX*). Inherited an *de novo ARX* mutations, including missense mutations and in frame duplications/insertions leading to expansions of polyalanine tracts in ARX, were found in nine familial an one sporadic case of MR (Bienvenu et al. 2002). In total,

Bienvenu et al., (Bienvenu et al. 2002) identified mutations in *ARX* in ten unrelated MRX families (7 out of the 9 families linked to Xp22.1, 2 out of 148 small families and 1 out of the 40 sporadic cases). Almost all available families with genetic intervals encompassing *ARX* were found to be mutated. These findings are interesting *per se* when compared with the very rare mutations (found in 1 to three families) that have been reported for most of the other known genes involved in MRX. In addition to the involvement of ARX in nonspecific MR, further data reported by Strømme et al. (Strømme et al. 2002) who have identified mutations in *ARX* in families with syndromic forms of mental retardation. These syndromes include: (1) X-linked West syndrome (WS) characterized by the triad of infantile spasms, chaotic electroencephalogram (EEG) patterns termed hypsarrythmia and mental retardation (ISSX, MIM 308350), (2) Partington syndrome (PRTS, MIM 309510) characterized by MR and dystonic movements of the hands, (3) MR associated with myoclonic epilepsy and spasticity. Phenotype/genotype data concerning ARX are particularly striking and uncommon. The spectrum of phenotypes associated with the identical recurrent duplication of the 24 bp of exon 2, predicted to cause an expansion of a polyalanine tract from 12 to 20 alanines, include nonspecific forms of mental retardation, West syndrome and Partington syndrome. Understanding the mechanisms underlying this clinical heterogeneity resulting from the same mutation is a difficult and challenging issue. One potential hypothesis to explain this phenotypic heterogeneity could be differences in genetic and environmental backgrounds which are obviously specific to each family.

In contrast to other genes involved in XLMR, *ARX* expression is specific to thetelencephalon and ventral thalamus. Notably there is an absence of expression in the cerebellum throughout development and also in adult (Bienvenu et al. 2002). These absence of detectable brain malformations in patients suggests that ARX may have an essential role, in mature neurons, required for the development of cognitive abilities.

## Molecular and cellular mechanisms underlying MRX

The genetic complexity underlying cognitive function appears enormous: Recent advances in genetics represent an important beginning in our efforts to understand the pathophysiology of MR. Delineation of the monogenic causes of MR and their molecular and cellular consequences will provide insights into the mechanisms required for normal development of cognitive functions in humans. The recent breakthroughs suggest that dysfunction of genes required for establishment, stabilisation and remodelling of connections between neuronal cells plays a major role in MRX (Chelly 1999; Luo 2000). Underlying this are defects in a variety of cellular processes, such as chromatin remodelling, gene expression, signal transduction and cytoskeletal dynamics.

It is interesting to point out that morphological abnormalities in dendritic spines were reported in patients with fragile X syndrome and in the corresponding mouse model deficient for the FMR1 gene (Comery et al. 1997). MRX gene homologues in mouse share similar temporal and spatial profiles of expression and, interestingly, the highest levels of expression are observed in the hippocampus, which is known to play an important role in learning and memory (J. Chelly et al., unpublished). Expression in mouse olfactory structures may correlate with the predominant role of olfactory perception in behavioural development in lower mammals. Finally, the absence of brain imaging abnormalities in MRX patients indicate, on the other hand, that nonspecific MR genes are not required for major aspects of brain development, such as layering of the cortex, organisation of the brain, neuronal migration and the initial steps of neuronal differentiation.

## Prospects for autosomal MR genes

While the excess of MR males has focused interest on XLMR, there is no reason to think that there is a higher density of genes involved in cognitive function on the X chromosome than on autosomes. Autosomal MR genes are therefore a target for further investigation. For autosomal recessive forms, in addition to the focus on syndromic MR (such as the numerous loci that are associated with MR and microcephaly; Pattison et al. 2000), investigation of large consanguineous families with nonspecific MR may lead to some success. Furthermore, if founder effects can be documented in some populations, this will help to restrict the size of the candidate region.

The situation is more difficult for autosomal dominant traits, because severe dominant MR can only occur through new mutations, or where there is high variability in expression of the phenotype (often seen in autosomal dominant traits). These situations render linkage analysis almost impossible in the absence of additional and specific clinical features. Even the characterization of genes at translocation breakpoints, or within large deletions associated with MR, will be difficult to validate by identification of mutations in patients without such rearrangements. Nevertheless, one should recall the important recent findings that telomeric chromosomal rearrangements (unbalanced translocations or deletions) might account for 5–7% of severe MR (in general associated with dysmorphic features or other clinical manifestations). Up to 50% of these rearrangements correspond to unbalanced translocations causing monosomies or trisomies of gene-rich subtelomeric regions that are inherited from a clinically normal parent carrying the balanced translocation. The latter cases have a high recurrence risk in the family (Knight et al. 1999).

## Diagnostic applications

The identification of the causing mutation in one of the MR genes is necessary for accurate diagnosis and counselling in a family where a genetic form of mental retardation may be suspected (in the absence of other, unambiguous, clinical criteria). For example, the identification of the expansion mutation responsible for the fragile X MR syndrome was very quickly implemented in all countries that could afford good genetic services. Most laboratories use the test broadly for

MR cases, with the result that the hit rate for positive diagnoses is rather low, in the 2–5% range (depending on whether some clinical preselection is implemented before testing). It should be noted that this molecular test will miss all conventional mutations in the FMR1 gene (apart from deletions affecting exon 1) that would however result in the same clinical outcome (see below).

For XLMR syndromes such as Coffin-Lowry or ATRX, the situation is more difficult. This is because there is a very wide diversity of mutations that can occur almost anywhere in these large genes, If a cell line is not available to do mutation screening at the mRNA level, it is necessary to screen the gene exon by exon. The cost of testing is such that the selection of patients by a competent clinical geneticist is mandatory to reach a reasonable hit rate for positive diagnoses (~35% for Coffin-Lowry syndrome; Delaunoy et al. 2001). Of course, such clinical screening will miss cases that do not fit the major clinical criteria. For instance, the RSK2 mutation that is associated with mild MR and no other clinical signs was found because it occurred in a very large family where the trait had been well mapped to Xp22 but one would never have searched in sporadic cases, or even in small families with two or three affected cases, with a similar phenotype (Merienne et al. 1999). A protein-based test (especially if it tests the function and not only the presence of the protein) may therefore have an advantage. However, it would be limited to target proteins present in easily sampled tissues (such as blood, hair and buccal epithelium; Willemsen et al. 1999; Merienne et al. 1998).

At present, diagnostic testing for the known nonspecific XLMR genes cannot be implemented in sporadic cases of male MR (or even if two males are affected in the same family), for technical and economical reasons. The average X-linked desease has an incidence of ~1 in 30,000–100,000 males (i.e., about 10-fold lower than that of the fragile X syndrome), and a similar frequency is likely for most XLMR genes.Higher incidences, such as those for Duchenne muscular dystrophy or Hemophilia A, are accounted for by the combination of a very large gene and the presence of one or several hot spots of mutation (indeed, the high frequency of the Rett syndrome appears to be due to the presence of hot spots for point mutations; see Table 3). Given the 1% incidence of MR in males, one can therefore deduce that the chance of observing a mutation in a given gene in a sporadic MR male patient will be 1 to 3 per 1000 (see Table 4). Given the average gene size of 15 to 20 exons, this will translate into one mutation observed for ~10,000 exons tested. There is no way, given the present technologies, that a laboratory would engage in such a search. Even limiting the search to probable X-linked cases (two affected brothers, or maternal uncle and nephew) would probably not increase efficiency by more than 4- or 5-fold. This explains why, for all nonspecific MR genes identified (with the exception of FMR2/FRAXE expansions), mutations have been found up to now in only two or three families. This also accounts for the striking deficit in FMR1 point mutations (Table 3), as the chances of observing such a mutation in one of the 15 exons is very low (Gronskov et al. 1998) because of the relatively poor specificity of the clinical phenotype elicited by loss of FMR1 function. By contrast, mutations have been identified in 86 Coffin-Lowry syndrome families (Delaunoy et al. 2001) and, even more strinkingly, 367 mutations have been reported for the MECP2 gene in Rett syndrome patients in less than two years (Bourdon et al. 2001; Laccone et al. 2001; Table 3).

**Table 3.** Mutation pattern in some XLMR genes

---

Fragile X/FMR1 (1991): 17 exons, 631 aa
   3 point mutations/1 missense; 12 deletions (in 15 families)
   CGG expansion: > 99% of families

Borderline XLMR with abnormal behavior / MAOA (1993): 15 exons, 527 aa
   1 nonsense mutation (in 1 family) (and few deletions in contiguous gene syndrome
   including Norrie disease gene)

Adrenoleukodystrophy / ALD (1993): 10 exons, 745 aa
   192 point mutations / 106 missense (55%); 12 deletions (in 335 families)
   one recurrent mutation accounts for 12% of patients

ATRX syndrome (1995): 35 exons, 2492 aa
   42 point mutations / 31 missense (74%), 1 deletion (in 62 families)
   one recurrent mutation accounts for 26% of patients

Coffin-Lowry syndrome / RSK2 (1996): 22 exons, 740 aa
   69 point mutations / 27 missnse (39%), 2 deletions (in 86 families)
   no highly recurrent mutation

X-linked lissencephaly/subcortical heterotopia / DCX (1998): 6 coding exons, 360 aa
   53 point mutations / 33 missense (62%), 1 deletion (in 77 families)
   4 recurrent mutations account for 27% of patients

Rett syndrome/MECP2 (1999): 3 exons, 486 aa
   61 point mutations / 28 missense (46%), 36 10-500 bp deletions (in 367 independent
   female patients): 7 recurrent mutations account for 63% of patients, and a hot spot of
   deletion for 10%

---

**Table 4.** Estimated frequencies of chromosomal and X-linked causes of mental retardation in males. The frequency values are very approximate, apart from those in bold. The actual frequency of detection of telomere deletions or fragile X mutations in MR male patients depends on the extent of preselection based on clinical phenotype.

| | Frequency in | |
| --- | --- | --- |
| | Male population | MR male patients |
| Moderate/severe MR | 1% | |
|    Chromosomal | 1/600 | 15% |
|    Telomere deletions | (1/2500–5000) | 2–4% |
| Fragile X | 1/4000–5000 | 2–2,5% |
| MECP2 | 1/6000–10000 | 1–1,5% |
| MRX nonspecific | | |
|    30–50 genes, each | 1/30000–1/100000 | 1/300 to 1/1000 |
| | $\sum$1/1000–2000 | $\sum$5–10% |
| Syndromic MRX | | |
|    30 genes | $\sum$1/1500–2000 | $\sum$5 7% |
| Total XLMR | 1.3–2.1/1000 | 13–21% |

In the latter case, the clinical features ar so specific that mutations are found in 70–80% of tested cases. If the 2% incidence of MECP2 missense mutations in male MR (Couvert et al. 2001) is confirmed in further studies, this gene will constitute an important diagnostic target in this patient population, especially as it is a relatively small gene (three exons).

If indeed there are 30–50 nonspecific MR genes (Chiurazzi et al. 2001; Chelly 1999) their cumulative incidence might account for 5–10% of MR in males, an important contribution to this medical problem (Table 4). The availability of a nearly complete X chromosome sequence, and the use of high throughput sequencing to screen for mutations in all genes mapping to the candidate regions in MRX families, should lead to identification of most MRX genes in the coming years. Detection of the mutations in these genes will be necessary if one wants to provide answers to parents of MR children about the cause of the problem they face, the risk of recurrence in the family and the possibility of prenatal diagnosis. How can this be achieved? One possibility may be through methodological improvements in mutation screening using automation and/or high density array strategies, which would decrease the cost of testing. However, as they stand, oligonucleotide arrays for mutation detection have a prohibitive cost, and many arrays would be required to cover the full range of XLMR genes (Hacia and Collins 1999). Another alternative is the development of protein-based assays and their implementation in an array format where many patients cell samples would be processed in parallel (Moch et al. 1999). However, as discussed above, this can only be applied to genes expressed in cells that can be sampled. Meanwhile, one could perhaps envisage a coordinated international effort to sequence systematically these MRX genes in a large number of families with demonstrated or possible XLMR (at least 1000 such families would be required). This would allow an assessment of the numerical contribution of each gene to XLMR and might identify mutations hot spots, which could be screened for mutations first in MR cases.

Finally, it should be stressed that genetic counselling and prenatal diagnosis related to mental handicap raise sensitive ethical issues, especially for the milder forms (such as for FRAXE, or for carrier females of a full fragile X mutation). Genetic diseases causing dysfunction of organs other than the brain can be documented using well-established laboratory or imaging techniques. Assessment of cognitive function is more complex, and performance can be subject to profound social and environmental factors in the family and the school. The level of expectation with respect to intellectual performance also depends on the family and on the type of society of which it is part. The controversies regarding the possible genetic basis for differences in IQ indicate that this domain is fraught with preconceptions and dogma.

## Acknowledgements

Research on XLMR in the authors`laboratories is supported by INSERM, EEC (grant QLG2-CT-1999-791) and Fondation pour la Recherche Médicale (JC and JLM), by funds from Association Française syndrome de Rett, AFM and Fonda-

tion Jerome Lejeune (to JC), and by FRAXA foundation, CNRS and the Hôpital
Universitaire de Strasbourg (HUS) (JLM).

## References

Allen KM, Gleeson JG, Bagrodia S, Partington MW, MacMillan JC, Cerione RA, Mulley JC, Walsh
    CA (1998) PAK3 mutation in nonsyndromic X-linked mental retardation. Nat Genet 20: 25–30
Amir R, Zoghbi H (2000) Rett syndrome: Methyl-CpG-binding protein 2 mutations and pheno-
    type-genotype correlations. Am J Med Genet (Sem Med Genet) 97: 147–152
Amir RE, Van den Veyver IB, Wan M, Tran CQ, Francke U, Zoghbi HY (1999) Rett syndrome is
    caused by mutations in X-linked MECP2, encoding methyl-CpG-binding protein 2. Nat Genet
    23: 185–188
Bagrodia S, Derijard B, Davis RJ, Cerione RA (1995) Cdc42 and PAK-mediated signaling leads to
    Jun kinase and p38 mitogen-activated protein kinase activation. J Biol Chem 270: 27995–27998
Bardoni B, Mandel JL, Fisch GS (2000) FMR1 gene and Fragile X syndrome. Am J Med Genet (Sem
    Med Genet) 97: 153–163
Berube NG, Smeenk CA, Picketts DJ Cell cycle-dependent phosphorylation of the ATRX protein
    correlates with changes in nuclear matrix and chromatin association (2000) Human Mol Genet
    9: 539–547
Bienvenu T, des Portes V, Martin AS, McDonell N, Billuart P, Carrié A, Vinet M-C, Couvert P,
    Toniolo D, Ropers HH, Moraine C, van Bokhoven H, Frijns J-P, Kahn A, Beldjord C, Chelly J
    (1998) Non-specific X-linked semidominant mental retardation by mutations in a Rab GDP-
    dissociation inhibitor. Hum Mol Genet 7: 1311–1315
Bienvenu T, des Portes V, McDonell N, Carrie A, Zemni R, Couvert P, Ropers HH, Moraine C, van
    Bokhoven H, Fryns JP, Allen K, Walsh CA, Boue J, Kahn A, Chelly J, Beldjord C (2000) Missense
    mutation in PAK3, R67C, causes X-linked nonspecific mental retardation In process Citation.
Bienvenu T, Poirier K, Friocourt G, Bahi N, Beaumont D, Fauchereau F, Ben Jeema L, Zemni R,
    Vinet MC, Francis F, Couvert P, Gomot M, Moraine C, van Bokhoven H, Kalscheuer V, Frints
    S, Gecz J, Ohzaki K, Chaabouni H, Fryns JP, desPortes V, Beldjord, C, Chelly J. ARX a Novel
    prd-class-homeobox Gene Highly Expressed in the Telencephalon, is Mutated in X-Linked
    Mental Retardation. Hum Mol Genet 2002; 11: 1–11.
Billuart P, Bienvenu T, Ronce N, Desportes V, Vinet MC, Zemni R, Crollius HR, Carrie A,
    Fauchereau F, Cherry M, Briault S, Hamel B, Fryns JP, Beldjord C, Kahn A, Moraine C, Chelly
    J (1998) Oligophrenin-1 encodes a rhoGAP protein involved in X-linked mental retardation.
    Nature 392: 923–926
Bourdon V, Philippe C, Labrune O, Amsallem D, Arnould C, Jonveaux P (2001) A detailed analysis
    of the MECP2 gene: prevalence of recurrent mutations and gross DNA rearrangements in Rett
    syndrome patients. Human Genet 108: 43–50
Cardoso C, Timsit S, Villard L, Khrestchatisky M, Fontes M, Colleaux L (1998) Specific interaction
    between the XNP/ATR-X gene product and the SET domain of the human EZH2 protein.
    Human Mol Genet 7: 679–684
Carrié A, Jun L, Bienvenu T, Vinet M-C, McDonell N, Couvert P, Zemni R, Cardona A, van
    Buggenhout G, Frints S, Hamel B, Moraine C, Ropers HH, Strom T, Howell GR, Whittaker A,
    Ross MT, Kahn A, Fryns J-P, Beldjord C,Marynen P, Chelly J (1999) A new member of the IL-1
    receptor family highly expressed in hippocampus and involved in X-linked mental retardation.
    Nat Genet 23: 25–31
Chelly J (1999) Breakthroughs in molecular and cellular mechanisms underlying X-linked mental
    retardation. Human Mol Genet 8: 1833–1838
Chen RZ, Akbarian S, Tudor M, Jaenisch R (2001) Deficiency of methyl-CpG binding protein-2
    in CNS neurons results in a Rett-like phenotype in mice. Nat Genet 27: 327–331
Chiurazzi P, Hamel BC, Neri G XLMR genes: update 2000. Eur J Human Genet 9: 71–81

Comery TA, Harris JB, Willems PJ, Oostra BA, Irwin SA, Weiler IJ, Greenough WT (1997) Abnormal dendritic spines in Fragile X knockout mice: Maturation and pruning deficits. Proc Nat Acad Sci USA 94: 5401–5404

Couvert P, Bienvenu T, Aquaviva C, Poirier K, Moraine C, Gendrot C, Verloes A, Andres C, Le Fevre AC, Souville I, Steffann J, des Portes V, Ropers HH, Yntema HG, Fryns JP, Briault S, Chelly J, Cherif B (2001) MECP2 is highly mutated in X-linked mental retardation. Human Mol Genet 10: 941–946

D´Adamo P, Menegon A, Lo Nigro C, Grasso M, Gulisano M, Tamanini F, Bienvenu T, Gedeon AK, Oostra B, Wu SK, Tandon A, Valtorta F, Balch WE, Chelly J, Toniolo D (1998) Mutations in GDI1 are responsible for x-linked non-specific mental retardation. Nat Genet 19: 134–139

De Cesare D, Jacquot S, Hanauer A, Sassone-Corsi P (1998) Rsk-2 activity is necessary for epidermal growth factor-induced phosphorylation of CREB protein and transcription of c-fos gene. Proc Natl Acad Sci USA 95: 12202–12207

Delaunoy J, Abidi F, Zeniou M, Jacquot S, Merienne K, Pannetier S, Schmitt M, Schwartz C, Hanauer A (2001) Mutations in the X-linked RSK2 gene (RPS6KA3) in patients with Coffin-Lowry syndrome. Human Mutat 17: 103–116

des Portes V, Beldjord C, Chelly J, Hamel B, Kremer H, Smits A, van Bokhoven H, Ropers HH, Claes S, Fryns J-P, Ronce N, Gendrot C, Toutain A, Raynaud M, Moraine C (1999) X- linked nonspecific mental retardation (MRX) linkage studies in 25 unrelated families. Am J Med Genet 85: 263–265

Donnai D, Kamiloff-Smith A (2000) Williams syndrome: from genotype through to the cognitive phenotype. Am J Med Genet 97: 164–171

Gecz J, Gedeon AK, Sutherland GR, Mulley JC (1996) Identification of the gene FMR2, associated with FRAXE mental retardation. Nat Genet 13: 105–108

Gecz J, Bielby S, Sutherland GR, Mulley JC (1997) Gene structure and subcellular localization of FMR2, a member of a new family of putative transcription activators. Genomics 44: 201–213

Geppert M, Bolshakov VY, Siegelbaum SA, Takei K, De Camilli P, Hammer RE, Sudhof TC (1994) The role of Rab3A in neurotransmitter release. Nature 369:493–497

Geppert M, Sudhof TC (1998) RAB3 and synaptotagmin: the yin and yang of synaptic membrane fusion. Annu Rev Neurosci 21: 75–95

Gibbons RJ, Higgs DR (2000) Molecular-clinical spectrum of the ATR-X syndrome. Am J Med Genet 97: 204–212

Gibbons RJ, Picketts DJ, Villard L, Higgs DR (1995) Mutations in a putative global transcriptional regulator cause X-linked mental retardation with alpha-thalassemia (ATR-X syndrome). Cell 80:837–845

Gibbons RJ, Bachoo S, Picketts DJ, Aftimo S, Asenbauer B, Bergoffen JA, Berry SA, Dahl N, Fryer A, Keppler K,Kurosawa K, Levin ML, Masuno M, Neri G, Pierpont ME, Slaney SF, Higgs DR (1997) Mutations in transcriptional regulator ATRX establish the functional significance of a PHD-like domain. Nat Genet 17: 146–148

Gibbons RJ, McDowell TL, Raman S, O´Rourke DM, Garrick D, Ayyub H, Higgs DR (2000) Mutations in ATRX, encoding a SWI/SNF-like protein, cause diverse changes in the pattern of DNA methylation. Nat genet 24: 368–371

Gronskov K, Hallberg A, Brondum-Nielsen K (1998) Mutational analysis of the FMR1 gene in 118 mentally retarded males suspected of fragile X syndrome: pathways for which proteins and biochemical partners involved are not fully identified

Gu Y, Shen Y, Gibbs RA, Nelson DL (1996) Identification of FMR2, a novel gene associated with the FRAXE CCG repeat and CpG island. Nat Genet 13: 109–113

Guerrini R, Shanahan JL, Carrozzo R, Bonanni P, Higgs DR, Gibbons RJ (2000) A nonsense mutation of the ATRX gene causing mild mental retardation and epilepsy. Ann Neurol 47: 117–121

Guy J, Hendrich B, Holmes M, Martin JE, Bird A (2001) A mouse Mecp2-null mutation causes neurological symptoms that mimic Rett syndrome. Nat Genet 27: 322–326

Hacia JG, Collins FS (1999) Mutational analysis using oligonucleotide microarrays. J Med Genet 36: 730–736

Hagermann RJ, Cronister A (1996) Fragile X syndrome: diagnosis, treatment and research. Johns Hopkins University Press, Baltimore, pp. 3–68

Imbert G, Feng Y, Nelson D, Warren ST, Mandel JL (1998) FMR1 and mutations in Fragile X syndrome: molecular biology, biochemistry, and absence of prevalent mutations [see comments]. Human Genet 102: 440–445

Ishizaki H, Miyoshi J, Kamiya H, Togawa A, Tanaka M, Sasaki T, Endo K, Mizoguchi A, Ozawa S, Takai Y (2000) Role of rab GDP dissociation inhibitor alpha in regulating plasticity of hippocampal neurotransmission. Proc Natl Acad Sci USA 97: 11587–11592

Knight SJ, Voelckel MA, Hirst MC, Flannery AV, Moncla A, Davies KE (1994) Triplet repeat expansion at the FRAXE locus and X-linked mild mental handicap. Am J Hum Genet 55:81–86

Knight SJ, Regan R, Nicod A, Horslea SW, Kearney L, Homfray T, W, Winter RM, Bolton P, Flint J (1999) Subtle chromosomal rearrangements in children with unexplained mental retardation. Lancet 354: 1676.1681

Knudsen J, Jensen MV, Hansen JK, Faergeman NJ, Neergaard TB, Gaigg B. Role of acylCoA binding protein in acylCoA transport, metabolism and cell signaling. Mol Cell Biochem 1999; 192: 95–103.

Kutsche K, Yntema H, Brandt A, Jantke I, Nothwang HG, Orth U, Boavida MG, David D, Chelly J, Fryns J-P, Moraine C, Ropers H-H, van Bokhoven H, Gal A (2000) Mutations in ARHGEF6, encoding a guanine nucleotide exchange factor for Rho GTPases, in patients with X-linked mental retardation. Nat Genet 26: 247–250

Laccone F, Huppke P, Hanefeld F, Meins M (2001) Mutation spectrum in patients with Rett syndrome in the German population: Evidence of hot spot regions. Human Mutat (2000) 17: 183–190

LeDouarin B, Nielsen AL, Garnier JM, Ichinose H, Jeanmougin F, Losson R, Chambon P (1996) A possible involvement of TIF1 and TIF1 in epigenetic control of transcription by nuclear receptors. EMBO J 15: 6701–6715

Lossi AM, Millan JM, Villard L, Orellana C, Cardoso C, Prieto F, Fontes M, Martinez (1999) Mutation of the XNP/ATR-X gene in a family with severe mental-retardation, spastic paraplegia and skewed pattern of X-inactivation – Demonstration that the mutation is involved in the inactivation BIAS. Am J Human Genet 65: 558 562

Luo L (2000) Rho GTPases in neuronal morphogenesis. Nat Rev Neurosci 1: 173–180

Maecker HT, Todd SC, Levy S (1997) The tetraspanin superfamily: Molecular facilitators FASEB J 11: 428–442

Manser E, Chong C, Zhao ZS, Leung T, Michael G, Hall C, Lim L (1995) Molecular cloning of a new member of the p21-Cdc42/Rac activated kinase (PAK) family. J Biol Chem 270: 25070–25078

McDowell TL, Gibbons RJ, Sutherland H, O´Rourke DM, Bickmore WA, Pombo A, Turley H, Gatter K, Picketts DJ, Buckle VJ, Chapman L, Rhodes D, Higgs DR (1999) Localization of a putative transcriptional regulator (ATRX) at pericentromeric heterochromatin and the short arms of acrocentric chromosomes. PNAS 96: 13983–13988

McLaren, Bryson SE (1987) Review of recent epidemiological studies of mental retardation: prevalence, associated disorders, and etiology. Am J Ment Retard 92: 243–254

Meloni I, Bruttini M, Longo I, Mari F, Rizzolio F, D´Adamo P, Denvriendt K, Fryos JP, Toniolo D, Renieri A (2000) A mutation in the rett syndrome gene, MECP2, causes X-linked mental retardation and progressive spasticity in males. Am J Human Gen 67: 982–985

Melonie I, Muscettola M, Raynaud M, Longo I, Bruttini M, Moizard MP, Gomot M, Chelly J, des Portes V, Fryns JP, Ropers HH, Magi B, Bellan C, Volpi N, Yntema HG, Lewis SE, Schaffer JE, Renieri A. FACL4, encoding fatty acid-CoA ligase 4, is mutated in nonspecific X-linked mental retardation. Nat Genet 2002; 30: 436–440.

Merienne K, Jaquot S, Trivier E, Pannetier S, Rossi A, Schinzel A, Castellan C, Kress W, Hanauer A (1998) Rapid immunoblot and kinase assay tests for a syndromal form of X-linked mental retardation: Coffin-Lowry syndrome. J Med Genet 35: 890–894

Merienne K, Jacquot S, Pannetier S, Zeniou M, Bankier A, Gecz J, Mandel JL, Mulley J, Sassone-Corsi P, Hanauer A. (1999) A missense mutation in RPS6KA3 (RSK2) responsible for nonspecific mental retardation. Nat Genet 22: 13–14

Miller WJ, Skinner JA, Foss GS, Davies KE (2000) Localization of the fragile X mental retardation 2 (FMR2)protein in mammalian brain. Eur J Neurosci 12: 381–384

Moch H, Schraml P, Bubendorf L, Mirlacher M, Kononen J, Gasser T, Mihatsch MJ, Kallioniemi OP, Sauter G (1999) High-throughput tissue microarray analysis to evaluate genes uncovered by cDNA microarray screening in renal cell carcinoma. Am J Pathol 154:981–986

Moser HW, Ramey CT, Leonhard CO (1990) Mental retardation. In: Principles and practice of medical genetics. Vol. 1. Emery AEH, Rimoin DL (eds) London – New York: Churchill Livingstone pp 495–511

Orrico A, Lam CW, Galli L, Dotti MT, Hayek G, Tong SF, Poon PMK, Zappella M, Federico A, Sorrentino V (2000) MECP2 mutation in male patients with non-specific X-linked mental retardation. FEBS Lett 481: 285–288

Pattison L, Crow YJ, Deeble VJ, Jackson AP, Jafri H, Rashid Y, Roberts E, Woods CG (2000) A fifth locus for primary autosomal recessive microcephaly maps to chromosome 1 q31. Am Human Genet 67: 1578–1580

Robertson KD, Wolffe AP (2000) DNA methylation in health and disease. Nat Rev Genet 1: 11–19

Sassone-Corsi P, Mizzen CM, Cheung P, Crosio C, Monaco M, Jacquot S, Hanauer A, Allis CD (1999) Requirement of Rsk-2 for Epidermal Growth Factor-activated phosphorylation of histone H3. Science 285: 886–891

Shaffer LG, Ledbetter DH, Lupski JR (2001) Molecular cytogenetics of contiguous gene syndromes: mechanisms and consequences of gene dosage imbalance. In: The metabolic and molecular bases of inherited disease. Vol. 1.Scriver CR, Beaudet AL, Sly WS, Valle D (eds) McGraw-Hill, New York, pp 1291–1326

Sirianni N, Naidu S, Pereira J, Pillotto RF, Hoffman EP (1998) Rett syndrome: confirmation of X-linked dominant inheritance, and localization of the gene to Xq28. Am J Human Genet 63: 1552–1558

Stevenson RE (2000) Splitting and lumping in the nosology of XLMR. Am J med Genet 97: 174–182

Stevenson RE, Schwartz CE, Schroer RJ (2000) X-linked mental retardation.Oxford: Oxford University Press

Strømme P, Mangelsdorf ME, Shaw MA, Lower KM, Lewis SM, Bruyere H, Lutcherath V, Gedeon AK, Wallace RH, Scheffer IE, Turner G, Partington M, Frints SG, Fryns JP, Sutherland GR, Mulley JC, Gecz J. Mutations in the human ortholog of Aristaless cause X-linked mental retardation and epilepsy. Nat Genet 2002; 30: 441–445.

Takagi S, Fujikawa K, Imai T, Fukuhara N, Fukudome K, Minegishi M, Tsuchiya S, Konno T, Hinuma Y, Yoshie O (1995) Identification of a highly specific surface marker of T-cell acute lymphoblastic leukemia and neuroblastoma as a new member of the transmembrane 4 superfamily. Intl J Cancer 61: 706–715

Trivier E, De Cesare D, Jacquot S, Pannetier S, Zackai E, Young I, Mandel JL, Sassone-Corsi P, Hanauer A. (1996) Mutations in the kinase Rsk-2 associated with Coffin-Lowry syndrome. Nature 384: 567–570

Villard L, Gecz J, Mattei JF, Fontes M, Saugier-Veber P, Munnich A, Lyonnet S (1996) XNP mutation in a large family with Juberg-Marsidi syndrome. Nat Genet 12: 359–360

Villard L, Fontes M, Ades LC, Gecz J (2000) Identification of a mutation in the XNP/ATR-X gene in a family reported as Smith-Fineman-Myers syndrome. Am J Med Genet 91: 83–85

Willemsen R, Anar B, De Diego Otero Y, de Vries BBA, Hilhorst-Hofstee Y, Smits A, van Looveren E, Willems PJ, Galjaard H, Oostra BA (1999) Noninvasive test for fragile X syndrome, using hair root analysis. Am J Human Genet 65: 98–103

Yntema HG, Van den Helm B, Kissing J, Van Duijnhoven G, Popelaars F, Chelly J, Moraine C, Frijns J-P, Hamel BCJ, Heilbronner H, Pander H-J, Brunner HG, Ropers HH, Cremers FPM, Van Bokhoven H (1999) A novel ribosomal S6-kinase (RSK4; RPS6KA6) is commonly deleted in patients with complex X-linked mental retardation. Genomics 62: 332–343

Youings SA, Murray A, Dennis N, Ennis S, Lewis C, McKechnie N, Pound A, Sharrock A, Jacobs P (2000) FRAXA and FRAXE: the results of a five year survey. J Med Genet 37: 415– 421

Zemni R, Bienvenu T, Vinet MC, Sefiani A, Carrie A, Billuart P, McDonell N, Couvert P, Francis F, Chafey P, Fauchereau F, Friocourt G, Portes Vd, Cardona A, Frints S, Meindl A, Brandau O, Ronce N, Moraine C, Bokhoven HV, Ropers HH, Sudbrak R, Kahn A, Fryns JP, Beldjord C (2000) A new gene involved in X-linked mental retardation identified by analysis of an X;2 balanced translocation. Nat Genet 24: 167–170

Zerial M, McBride H (2001) Rab proteins as membrane organizers. Nat Rev Mol Cell Biol 2:107–117

Warren ST, Wells RD (eds) In: Genetic instabilities and hereditary neurological diseases. San Diego: Academic Press, (2001), pp 27–53

# Gene Therapy: Medicine of the 21$^{st}$ Century

*I.M. Verma*[1]

## Introduction

The basic concept of gene therapy is disarmingly simple – introduce the gene, and its product should have the ability to cure a disease or slow down the progression of a disease. Encompassed within this definition are a number of different goals, including the treatment of both inherited and acquired disease. Treatment requires a technology capable of gene transfer in a wide variety of cells, tissues and whole organs. The therapeutic response may range from a total correction of symptoms to an improvement in the quality of life of a patient. Central to this notion is the ability to transfer genes, but the delivery vehicles needed to ferry genetic material into a cell still represent the "Achilles' heel" of gene therapy. An ideal vector should have the following attributes:

1) **Efficient and easy production:** High titer preparations of vector particles should be reproducibly available. The efficient transduction of cells within tissues is only possible if a sufficient number of infectious particles reaches the target cells. For the widespread use of viral vectors, facile production procedures have to be developed.
2) **Safety aspects:** The vector should neither be toxic to the target cells nor induce unwanted effects, including immunological reactions against the viral vector or its cargo. The latter carries not only the threat to eliminate the vector and/or the infected cells but may also lead to life-threatening complications, such as septic shock.
3) **Sustained and regulated transgene expression:** The gene delivered by the viral vector has to be expressed in a proper way. Permanent or even life-long expression of the therapeutic gene is desired only in a minority of diseases (e.g., treatment of hemophilia). Controlled expression of the transgene in a reversible manner would be highly desirable in many cases (e.g., gene therapy for insulin-dependent diabetes mellitus).
4) **Targeting of the viral vectors:** Preferential or exclusive transduction of specific cell types is very desirable.

[1]The Laboratory of Genetics, The Salk Institute, 10010 North Torrey Pines Road, La Jolla, CA 92037, 858-453-4100 x1462, fax: 858-558-7454, e-mail: verma@salk.edu

Mallet/Christen
Neurosciencess at the Postgenomic Era
© Springer-Verlag Berlin Heidelberg 2003

5) **Infection of dividing and non-dividing cells:** Since the majority of the cells in an adult human being are in a post-mitotic, non-dividing state, viral vectors should be able to efficiently transduce these cells.
6) **Site-specific integration:** Integration into the host genome at specific site(s) could enable us to repair genetic defects, such as mutations and deletions, by insertion of the correct sequences. Thus, it would no longer be necessary to replace defective gene expression by introducing foreign genes and cDNAs.

The vectors available now fall into two broad categories – the non-viral and viral vectors. The non-viral vectors are naked DNA delivered by injection and liposomes (cationic lipids mixed with nucleic acids; Li and Huang 2000). While nonviral vectors can be produced in relatively large amounts and are likely to present fewer toxic or immunological problems, presently they suffer from inefficient gene transfer. Furthermore expression of the foreign gene is transient. Given the need, in many diseased states, for sustained and in many cases high expression of the transgene, viral vectors are the most suitable vehicles for efficient gene delivery (Li and Huang 2000; Templeton and Lasic 2000).

## Current viral vectors

All viruses have a genetic component that is essential for further propagation. Viral vectors are derived from viruses by replacing these genetic components with the therapeutic gene. In a cell (typically called a packaging cell), the essential components can be provided in *trans*, which enables the viral vector to be packaged and to deliver genes to the target cell. But this is a dead-end infection, because the virus is lacking the essential components for viral propagation. Recombination between the vector and the viral genes encoding the essential components in the packaging cells can lead to the generation of infectious parental virus. Therefore, the removal and/or the separation of gene encoding viral components essential for viral propagation reduces the risk of generating infectious virus – a principle frequently used in gene therapy vector design.

The viral vectors can be divided into two general categories, integrating and non-integrating; the former holds the promise of life-long expression of the deficient gene product. At present there are three main vectors capable of integration in the recipient cells (retroviral, lentiviral, adeno-associated viral) and one vector type that is maintained as an episome (adenoviral).

### Retroviral Vectors

Vectors based on retroviruses were among the first to be designed and they have been important for the technical and conceptual development of viral vectors as a whole (Verma 1990; Miller 1992; Mulligan 1993; Anderson 1998). Retroviruses have three essential genes, which can be provided separately in packaging cells: *gag* encodes viral structural proteins; *pol* encodes reverse transcriptase/integrase; and *env* encodes viral envelope glycoprotein. Conceptually, the idea of separating *gag-pol* from *env* offered a packaging cell line where the chances of generating replication-competent retroviruses (RCR) were significantly reduced. Retroviral vectors also contributed towards the idea of changing the envelope protein to modify the range of target cells. Envelope proteins such as ecotropic (infection of only rodent cells), xenotropic (infection of most mammalian cells except rodent cells), amphotropic (infection of all mammalian cells) to pantropic (infection of cells from a variety of species) have been used (Danos and Mulligan 1988; Markowitz et al. 1988; Burns et al. 1993). Another conceptual breakthrough in generating safe and transcriptionally regulatable vectors was the development of self-inactivating (SIN) vectors (Yu et al. 1986) where the viral regulatory elements have been deleted. Therefore upon integration all viral promoter/enhancer activity is lost was and the transcription of the transgene is under the control of a heterologous promoter.

The use of retroviral vectors also led to the technological production, storage and distribution of commercial vectors on the scale required for human clinical trials. There are currently a wide variety of packaging cell lines and vectors with improved transduction efficiencies. They include novel features such as tissue-specific promoters, inducible promoters, internal ribosomal entry sites (IRES -to allow translation of multiple proteins from a single transcript) and *env* proteins with modified target specificity (Morgan and Anderson 1993; Friedmann 1999). The major limitation of retroviral vectors has been their inability to infect non-dividing cells. Therefore, tissues such as brain, eye, lungs, and pancreas are not amenable to direct *in vivo* gene delivery. Furthermore, upon transplantation in the host, transcription of the transgene is often extinguished (St. Louis and Verma 1988; Palmer et al. 1991). These two serious limitations have lead many scientists to explore vectors capable of infecting non-dividing cells as well as integrating into the host chromosome.

### Lentiviral vectors

Lentiviruses are complex retroviruses that have been named (*lenti*, latin for slow) according to the prototypic, slowly progressing neurologic disease in sheep caused by the maedi/visna virus (Joag et al. 1996). An important genetic difference between simple retroviruses and lentiviruses are regulatory (*tat* and *rev*) and auxillary genes (*vpr*, *vif*, *vpu*, and *nef*) that have important functions during the viral life cycle and viral pathogenesis (Coffin 1996; Hirsch and Curran 1996; Joag et al. 1996). An outstanding feature of lentiviruses is their ability to infect non-dividing, terminally differentiated mammalian cells, including lymphocytes and macrophages. This feature of lentiviruses makes them a very attractive tool for gene delivery (Naldini and Verma 1999; Trono 2000).

The first lentiviral vectors were derived from HIV-1 (Naldini et al. 1996; Poeschla et al. 1996; Reiser et al. 1996; Pfeifer et al. 2001a), the most extensively studied lentivirus. The HIV vector and packaging system are constantly evolving and serve as templates for the other lentiviral vectors. Apart from HIV-1, lentivirus vectors have been derived from HIV-2 (Poeschla et al. 1998a), feline immunodeficiency virus (FIV) (Poeschla et al. 1998b), equine infectious anemia virus (Olsen 1998), simian immunodeficiency virus (SIV) (Mangeot et al. 2000), and maedi/visna virus (Berkowitz et al. 2001). Most of the lentiviral vectors presently in use for gene therapy approaches are HIV-derived vectors; therefore, we will focus on these vectors. Similar to simple retroviruses, the *cis*- and *trans*-acting factors of lentiviruses can be separated while preserving their functions. The lentiviral packaging systems provide in *trans* the viral proteins that are required for the assembly of viral particles in the packaging cells. The vector constructs contain the viral *cis* elements, packaging sequences (<F128M>Y<F255M>), the Rev response element (RRE), and the transgene (Fig. 1).

HIV is able to infect non-dividing and terminally differentiated cells without requiring the disassembly of the nuclear membrane (Weinberg et al. 1991). The HIV preintegration nucleoprotein complex can enter the nucleus via nuclear localization signal-mediated, energy-dependent import (for references see Brown 1997).

The development of regulated lentiviral vectors has focused – so far – only on the tetracycyline-regulated system (Kafri et al. 2000) (see also above, "Retrovirus Vectors-Regulation of Transgene Expression"). With this regulatable lentiviral vector, tet-dependent induction of transgene expression (i.e., GFP) was demonstrated in vitro and in vivo (rat brain). Another important improvement of the lentiviral vectors is the inclusion of *cis*-acting transcriptional regulatory elements, such as the WPRE, which enhances transgene expression in the target cells. Incorporation of WPRE in HIV-derived vectors increases reporter gene (GFP and luciferase) expression 5- to 8-fold after transduction of both dividing and arrested 293T cells (Zufferey et al. 1999). Similar to vectors based on MLV, the WPRE has to be present within the transgene transcript in sense orientation and is placed 3′ of the transgene cDNA upstream of the 3` LTR (Fig. 1).

So far, most of the lentiviral vectors have been produced by transient transfection of packaging and vector plasmids. Using the highly transfectable 293T cells, one routinely obtains titers of $1 \times 10^9$-$1 \times 10^{10}$ IU/ml after transient transfection with the latest generations of packaging and vector constructs followed by concentration of the virus particles by ultracentrifugation. However, standardization of the virus production is not easily achieved with transient transfections, and each preparation should be tested for possible contaminations with RCRs. Thus, stable producer cell lines may be advantageous, especially if HIV-derived vectors are used in clinical trials.

The development of a packaging cell line for lentiviral vectors was hampered by the fact that VSV-G and some lentiviral proteins (Vpr, Gag and Tat) are toxic. Therefore, packaging cell lines were established that express the toxic components from inducible plasmids (Yu et al. 1996; Kafri et al. 1999) using tetracycline-regulated systems (see above). In addition, stable cell lines for the production of SIN vectors (Xu et al. 2001) and stable third generation lentiviral packaging systems have been generated (Klages et al. 2000). A few examples of transduction by lentiviral vectors are seen in Figure 2.

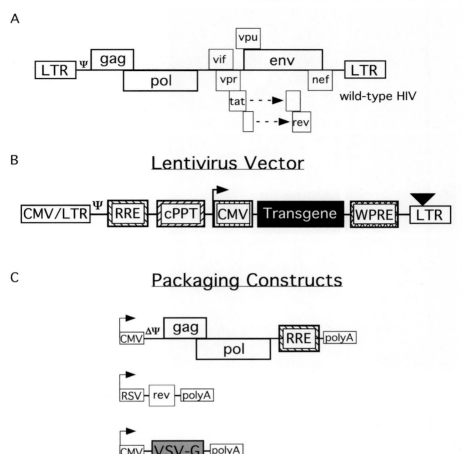

**Fig. 1.** Lentiviral vectors. (A) Schematic representation of the wild-type HIV provirus. (B) The latest generation of SIN-lentiviral vector constructs incorporates a central poly purine tract (cPPT) to enhance nuclear translocation of the vector in the target cell. In addition, a WPRE is included. Black triangle, SIN mutation; RRE, Rev response element; (C) Third generation, Tat-free, non-overlapping split genome packaging system. One plasmid codes for *gag* and *pol*, while *rev* is expressed in *trans* from another plasmid (Dull et al. 1998).

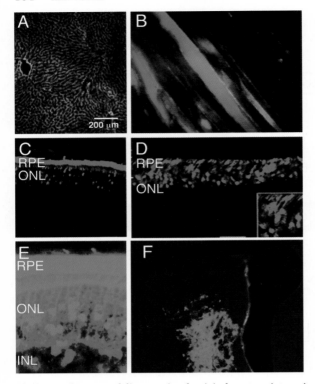

**Fig. 2.** In vivo gene delivery using lentiviral vectors into mice and rats. The gene for green fluorescent protein (GFP) was placed into lentiviral vectors, and these were injected into various tissues. The figure shows efficient gene transduction (green cells) to liver (A), and muscle (B). Different promoters driving transgene expression. The cytomegalovirus (CMV) promoter (C) or the rhodopsin promoter (D) driving GFP expression. Note the high level expression of GFP in the retinal pigment epithelium (RPE) with the CMV promoter and in photoreceptor cells with the rhodopsin promoter. A SIN lentivirus vector expresses under the CMV promoter in the retina (E) and in the brain (F). Regulated expression is illustrated with a lentiviral vector where the GFP gene is under the control of the tet regulatable system (see text).

## Adeno-associated viral vectors

Adeno-associated virus (AAV) is a small, non-pathogenic single-stranded DNA virus that has turned out to be a very efficient and useful delivery vehicle (Carter and Samulski 2000). The virus is a member of the dependoviruses and, as the name suggests, requires additional genes to replicate (Berns 1996). These genes are provided by adenovirus (hence the name, since AAV was found to be associated with this virus) or herpes virus. The AAV itself has two genes: *rep* coding for replication and integration functions of the virus, and *cap* coding for the structural components of the virus. *Rep* and *cap* are sandwiched between two inverted terminal repeats (ITRs) that define the beginning and the end of the virus and contain DNA sequences needed for packaging its genome into viral capsids.

The viral vector is produced by replacing the *rep* and *cap* genes with the therapeutic gene. The *rep* and *cap* gene products are produced *in trans* in the packaging cell, as are the adenoviral genes needed for replication of the virus (Fig. 2). AAV is capable of integrating in a site-specific location on chromosome 19, through the action of the *rep* protein. However, the vector, on entry into the target cell, does not code for *rep*, losing this highly desirable attribute (Kotin et al. 1990). Another interesting feature of AAV vectors with respect to chromosomal integration is their propensity for homologous recombination. The vector has been demonstrated to correct point mutations and deletions in selectable reporter genes integrated into the chromosome, although this occurs at a very low frequency (Inoue et al. 1999).

When tested in mice, dogs and monkeys, gene expression from AAV vectors is sustained in tissues with long-lived cells such as muscle, liver and brain (Carter and Samulski 2000). This long-term expression results from randomly integrated vectors and some vector DNA that persists as extra-chromosomal DNA. At present it is not clear what proportion of expression originates from integrated or extra-chromosomal DNA(Malik et al. 1997).

The main problem with AAV vectors is that the *rep* gene, and some of the adenoviral helper genes, are cytostatic and cytotoxic to the packaging cells, and so it has been difficult to scale up production of these viral vectors. No cell lines have been reported which stably produce high-titer AAV vectors carrying a therapeutic gene. Current clinical trials with AAV vectors rely on transient production systems, which may be suitable for proof of principle, but a more efficient production system is urgently required (Kay et al. 2000). Another problem is the coding capacity of the vector, which is restricted to less than 4 to 4.5 kb of nucleic acid. However, two groups have extended the packaging capacity of these vectors by using the observation that AAV genomes concatomerize after transduction. When two vectors, one encoding the first half and the other encoding the second half of a protein, were transduced into cells, head-to-tail stitching of the viral genomes resulted in the reconstitution of a functional gene, effectively increasing the size of the gene that can be delivered. It remains to be seen if concatomerized vectors will be stable and have sustained expression (Nakai et al. 2000 Yan et al. 2000).

### Adenoviral vectors

The adenoviruses are a family of DNA tumor viruses that cause benign respiratory tract infections in humans (Shenk 1996). The genome contains over a dozen genes; upon infection the virus remains episomal in the nucleus and can transduce genetic material into both dividing and non-dividing cells. Replication-defective recombinant adenoviral vectors can be generated at high-titers ($>10^{13}$ – $10^{14}$ particles/ml) by deleting a number of viral genes, including E1A, E1B, E3, E4 and E2A (Yeh and Perricaudet 1997). Most recently, "gut-less" adenoviral vectors (where all the viral genes are removed and provided *in trans*) have also been generated (Kochanek et al. 1996; Parks et al. 1996). All adenoviral vectors retain the ability to transduce very efficiently dividing and non-dividing cells and it is relatively easy to generate high-titer, commercial-grade recombinant vectors.

The challenge for adenoviral vectors concerns the length of expression of the transgene. All adenoviral vectors to date, with the exception of "gut-less" or fully deleted (data for which is still preliminary) (Morral et al. 1999) express the transgene in adult animals for only a short period of time (between 5 and 20 days post-infection) (Dai et al. 1995). In immuno-compromised animals, expression in slowly "turning-over" cells such as muscle cells and neurons is observed for long periods of time. It is now generally recognized that the short-term expression from recombinant adenoviral vectors is due to the immune response (see below). These vectors will, however, continue to be utilized in situations where high-level but transient expression of the foreign gene is required, for example in restenosis and cancer.

## Clinical applications

Worldwide, over 400 clinical gene therapy trials have been conducted or are underway, with enrollment of over 6,000 patients. Substantial numbers of these clinical trials (over 70%) are cancer related and often carried out on terminal patients. The most commonly used vectors are retroviral vectors based on MLV, which were the first viral vectors to be used in a gene therapy trial (Blaese et al. 1995). The target of this first clinical trial was the T-lymphocytes of two children suffering from severe combined immunodeficiency (SCID; Blaese et al. 1995). Unfortunately, a clear judgment of the success of this trial was not possible since the patients received supportive conventional therapy. However, this trial did set the stage for other gene therapy trials.

Gene transfer into multipotent hematopoietic stem cells has received much attention due to its relevance for a broad variety of human diseases, ranging from hematological disorders to cancer. In addition, it allows the use of *ex vivo* transduction protocols, thereby minimizing the exposure of the patient with viral particles. The use of retroviral vectors in this setting is hampered by the low frequency of gene delivery, since transduction by retrovirus vectors occurs only in cells that are replicating at the time of infection (Miller et al. 1990), and therefore the transduction of slowly or non-dividing stem cells and progenitors is inefficient. A number of growth factor combinations have been used to prestimulate hematopoietic stem/progenitor cells and have been shown to increase transduction efficacy (Nolta et al. 1995). The drawback of this approach is that exposure of progenitors to growth factors over several days markedly impairs their ability for long-term engraftment (Tisdale et al. 1998). However, a brief (<24 h) exposure to specific cytokines and stromal support allows engraftment with a high number (10–20%) of retrovirally modified cells.

Given the positive effect of bone marrow stroma on retroviral gene delivery to stem cells, attempts were made to mimic the bone marrow milieu by addition of purified extracellular matrix molecules. Certain fibronectin fragments (e.g., fibronectin CH296; Moritz et al. 1996) proved to significantly increase the gene transfer efficiency due to colocalization of retroviral particles and target cells (Hanenberg et al. 1996). This procedure resulted in a relatively high level (median 14%) of gene transfer in human CD34-positive cells as assayed by the number of transduced progenitor colonies (Abonour et al. 2000). Although a median engraftment level of 12% transduce cells was observed in human bone marrow one month after transplantation, the number of transduced cells fell to 5% over the next 11 months (Abonour et al. 2000). The combination of fibronectin and cytokine co-cultivation is a promising approach to retroviral transduction of stem cells and has resulted in an improved gene transfer to baboon marrow stem cells (Kiem et al. 1998) and was used for the correction of the SCID phenotype in two patients (Cavazzana-Calvo et al. 2000).

Among the clinical trials presently conducted in the U.S. that involve AAV, one has received much attention. In this phase I trial, patients suffering from hemophilia B were injected with an AAV vector carrying the cDNA for factor IX into the skeletal muscle (Kay et al. 2000). Preliminary data on three patients that received the starting dose of the dose-escalation study – i.e., $10^{11}$ I.U./kg – suggest that the transduced muscle cells expressed factor IX protein for at least two to three months. In addition, an 80% reduction of factor IX infusion was observed in one patient and was interpreted as evidence for a modest clinical response. However, the reduction in factor usage is a quite surprising finding, given the fact that the factor IX levels were not affected by the AAV-factor IX injection in the same patient. The fact that so far neither vector-related toxicity (at least at the initial dose) nor evidence for germline transmission of vector sequences was found in these patients is an important finding but has to be confirmed at higher, therapeutic vector doses. Another important question is the immune response of the patients against the vector. This studies by Chirmule et al. (1999) and Kay et al. (2000) and other studies showed a pre-existing humoral immunity against AAV in virtually all patients. In addition, administration of the AAV vector elicited a 10- to 1,000-fold increase in the neutralizing antibody titer (Kay et al. 2000). This boost of the humoral immunity against the vector is of major concern, especially if the AAV vectors have to be re-administered.

Two features of lentiviral vectors are of outstanding interest regarding their use in human beings: 1) the ability to transduce non-dividing cells in vivo, and 2) the ability to efficiently deliver large (~8 kb) and complex transgenes to the target cells and tissues.

HIV-derived lentiviral vectors have been shown to transduce a broad spectrum of non-dividing cells in vivo, such as neurons (Naldini et al. 1996), retinal cells (Miyoshi et al. 1998; Takahashi et al. 1999), muscle cells (Kafri et al. 1997), and hepatocytes (Kafri et al. 1997; Pfeifer et al. 2001b). An important finding is that lentiviral vectors can transduce human CD34$^+$ hematopoietic stem cells without cytokine prestimulation (Case et al. 1999; Miyoshi et al. 1999; Follenzi et al. 2000). The transduced CD34$^+$ cells provided long-term repopulation and were capable of engrafting and differentiation into multiple hematopoietic lineages after transplantation into NOD/SCID mice (Miyoshi et al. 1999). Since bone marrow stem cells can be transduced ex vivo, there is no requirement to use "live" virus in patients, making hematopoietic stem cells the most likely target for lentiviral gene therapy trials in humans.

## Perspectives

The young field of gene therapy promises major medical progress towards the cure of a broad spectrum of human diseases, ranging from immunological disorders to heart disease and cancer, and has, therefore, generated great hopes and great hype. The idea of using the genetic information obtained by sequencing of the human genome for the treatment of diseases is compelling. However, gene therapy will only be added to the daily-use therapeutic arsenal if scientists from many different disciplines participate and pull together as a team: Geneticists will have to identify target genes that contribute to specific diseases or that can influence the disease course. The task for the virologists will be to develop efficient and safe vectors that are able to deliver the genes of interest to the target cells and assure the proper expression of the transferred genetic material. Cell biologists will establish ways to facilitate the gene transfer, and identify stem cells that can be used to regenerate failing organs. Bioengineers will be needed to show the biologists how three-dimensional tissues and even whole organs may be generated in a test tube. Clinicians will carry out clinical trials with vectors optimized for the disease and the medical requirements of the patients.

Gene therapy has undergone extreme scrutiny in the recent past. It is our responsibility to assure the public that the patient´s welfare and health is the major goal. Strict adherence to the guidelines is incumbent on all scientists and investigators involved in a clinical trial. The gene therapy community will need to meet the challenge of new regulations and guidelines introduced by the NIH and FDA to ensure both the quality of clinical trials and protection of volunteers enrolled in the trials. We want to continue to participate and lead the nation in harnessing and providing the benefits of the unprecedented golden age of biomedical research.

## Acknowledgments

Most of the contents of this manuscript have previously been published. I would like to thank many collaborators, most notably, Alexander Pfeifer, Francesco Galimi, Tal Kafri, Didier Trono, Luigi Naldini, Hiroyuki Miyoshi, Nikunj Somia, and Fred Gage for their contributions that led to the development of lentiviral vectors. Dr. Verma is an American Cancer Society Professor of Molecular Biology. He is supported by grants from the NIH, the March of Dimes, the Lebensfeld Foundation, the Wayne and Gladys Valley Foundation, and the H.N. and Frances C. Berger Foundation.

# References

Abonour R, Williams DA, Einhorn L, Hall KM, Chen J, Coffman J, Traycoff CM, Bank A, Kato I, Ward M, Williams SD, Hromas R, Robertson MJ, Smith FO, Woo D, Mills B, Srour EF, Cornetta K (2000) Efficient retrovirus-mediated transfer of the multidrug resistance 1 gene into autologous human long-term repopulating hematopoietic stem cells. Nature Med 6: 652–658

Anderson WF (1998) Human gene therapy. Nature 392(Suppl): 25–30

Berkowitz RD, Plavec I, Veres G (2001) Gene transfer systems derived from visna virus: analysis of virus production and infectivity. Virology 279: 116–129

Berns KI (1996) *Parvoviridae:* the Viruses and their replication. In: Fields BN, Knipe DM, Howley PM (eds) Fields virology Vol. 2. Philadelphia, Lippincott-Raven Publishers, pp. 2173–2198

Blaese RM, Culver KW, Miller AD, Carter CS, Fleisher T, Clerici M, Shearer G, Chang L, Chiang Y, Tolstoshev P, Greenblatt JJ, Rosenberg A, Klein H, Berger M, Mullen A, Ramsey JW, Muul L, Morgan RA, French Anderson W (1995) T lymphocyte-directed gene therapy for ADA-SCID: initial trial results after 4 years. Science 270: 475–480

Brown PO (1997) Integration. In: Coffin JM, Hughes SH, Varmus HE (eds) Retroviruses. New York, Cold Spring Harbor Laboratory Press, 161–203

Burns JC, Friedmann T, Driever W, Burrascano M, Yee JK (1993) Vesicular stomatitis virus G glycoprotein pseudotyped retroviral vectors: concentration to very high titer and efficient gene transfer into mammalian and nonmammalian cells. Proc Natl Acad Sci USA 90: 8033–8037

Carter PJ, Samulski RJ (2000) Adeno-associated viral vectors as gene delivery vehicles. Intl J Mol Med 6: 17–27

Case SS, Price MA, Jordan CT, Yu XJ, Wang L, Bauer G, Haas DL, Xu D, Stripecke R, Naldini L, Kohn DB, Crooks GM (1999) Stable transduction of quiescent CD34(+)CD38(-) human hematopoietic cells by HIV-1-based lentiviral vectors. Proc Natl Acad Sci USA 96: 2988–2993

Cavazzana-Calvo M, Hacein-Bey S, de Saint Basile G, Gross F, Yvon E, Nusbaum P, Selz F, Hue C, Certain S, Casanova JL, Bousso P, Deist FL, Fischer A (2000) Gene therapy of human severe combined immunodeficiency (SCID)-X1 disease. Science 288: 669-672

Chirmule N, Propert K, Magosin S, Qian Y, Qian R, Wilson J (1999) Immune responses to adenovirus and adeno-associated virus in humans. Gene Ther 6: 1574–1583

Coffin JM (1996) Retroviridae: The viruses and their replication. In: Fields BN, Knipe DM, Howley PM (eds) Fields virology. Philadelphia, Lippincott-Raven Publishers, pp. 1767–1848

Dai Y, Schwarz EM, Gu D, Zhang WW, Sarvetnick N, Verma IM (1995) Cellular and humoral immune responses to adenoviral vectors containing factor IX gene: tolerization of factor IX and vector antigens allows for long-term expression. Proc Natl Acad Sci USA 92: 1401–1405

Danos O, Mulligan RC (1988) Safe and efficient generation of recombinant retroviruses with amphotropic and ecotropic host ranges. Proc Natl Acad Sci USA 85: 6460-6464

Dull T, Zufferey R, Kelly M, Mandel RJ, Nguyen M, Trono D, Naldini L (1998) A third-generation lentivirus vector with a conditional packaging system. J Virol 72: 8463-8471

Follenzi A, Ailles LE, Bakovic S, Geuna M, Naldini L (2000) Gene transfer by lentiviral vectors is limited by nuclear translocation and rescued by HIV-1 pol sequences. Nat Genet 25: 217–222

Friedmann T (ed) (1999) The development of human gene therapy. New York, Cold Spring Harbor Laboratory Press

Hanenberg H, Xiao XL, Dilloo D, Hashino K, Kato I, Williams DA (1996) Colocalization of retrovirus and target cells on specific fibronectin fragments increases genetic transduction of mammalian cells. Nature Med 2: 876–882

Hirsch MS, Curran J (1996) Human immunodeficiency viruses. In: Fields BN, Knipe DM, Howley (eds) Fields virology. Philadelphia, Lippincott-Raven Publishers, pp. 1953–1975

Inoue N, Hirata RK, Russell DW (1999) High-fidelity correction of mutations at multiple chromosomal positions by adeno-associated virus vectors. J Virol 73: 7376–7380

Joag SV, Stephens EB, Narayan O (1996) Lentiviruses. In: Fields BN, Knipe DM, Howley PM (eds) Fields virology Philadelphia, Lippincott-Raven Publishers, pp. 1977–1996

Kafri T, Blomer U, Peterson DA, Gage FH, Verma IM (1997) Sustained expression of genes delivered directly into liver and muscle by lentiviral vectors. Nature Genet 17: 314–317

Kafri T, van Praag H, Ouyang L, Gage FH, Verma IM (1999) A packaging cell line for lentivirus vectors. J Virol 73: 576–584

Kafri T, van Praag H, Gage FH, Verma IM (2000)Lentiviral vectors – regulated gene expression. Mol Ther 1: 516–521

Kay MA, Manno CS, Ragni MV, Larson PJ, Couto LB, McClelland A, Glader B, Chew AJ, Tai SJ, Herzog RW, Arruda V, Johnson F, Scallan C, Skarsgard E, Flake AW, High KA (2000) Evidence for gene transfer and expression of factor IX in haemophilia B patients treated with an AAV vector. Nat Genet 24: 257–261

Kiem HP, Andrews RG, Morris J, Peterson L, Heyward S, Allen JM, Rasko JE, Potter J, Miller AD (1998) Improved gene transfer into baboon marrow repopulating cells using recombinant human fibronectin fragment CH-296 in combination with interleukin-6, stem cell factor, FLT-3 ligand, and megakaryocyte growth and development factor. Blood 92: 1878–1886

Klages N, Zufferey R, Trono D (2000) A stable system for the high-titer production of multiply attenuated lentiviral vectors. Mol Ther 2: 170–176

Kochanek S, Clemens PR, Mitani K, Chen HH, Chan S, Caskey CT (1996) A new adenoviral vector: Replacement of all viral coding sequences with 28 kb of DNA independently expressing both full-length dystrophin and beta-galactosidase. Proc Natl Acad Sci USA 93: 5731–5736

Kotin RM, Siniscalco M, Samulski RJ, Zhu XD, Hunter L, Laughlin CA, McLaughlin S, Muzyczka N, Rocchi M, Berns KI (1990) Site-specific integration by adeno-associated virus. Proc Natl Acad Sci USA 87: 2211–2215

Li S, Huang L (2000) Nonviral gene therapy: promises and challenges. Gene Ther 7: 31–34

Malik P, McQuiston SA, Yu XJ, Pepper KA, Krall WJ, Podsakoff GM, Kurtzman GJ, Kohn DB (1997) Recombinant adeno-associated virus mediates a high level of gene transfer but less efficient integration in the K562 human hematopoietic cell line. J Virol 71: 1776–1783

Mangeot PE, Negre D, Dubois B, Winter AJ, Leissner P, Mehtali M, Kaiserlian D, Cosset FL, Darlix JL (2000) Development of minimal lentivirus vectors derived from simian immunodeficiency virus (SIVmac251) and their use for gene transfer into human dendritic cells. J Virol 74: 8307–8315

Markowitz D, Goff S, Bank A (1988) A safe packaging line for gene transfer: separating viral genes on two different plasmids. J Virol 62: 1120–1124

Miller AD (1992) Human gene therapy comes of age. Nature 357455–357460

Miller DG Adam MA, Miller AD (1990) Gene transfer by retrovirus vectors occurs only in cells that are actively replicating at the time of infection [published erratum appears in Mol Cell Biol 1992 Jan; 12: 433]. Mol Cell Biol 10: 4239–4242

Miyoshi H, Blomer U, Takahashi M, Gage FH, Verma IM (1998) Development of a self-inactivating lentivirus vector. J Virol 72: 8150–8157

Miyoshi H, Smith KA, Mosier DE, Verma IM, Torbett BE (1999) Transduction of human CD34+ cells that mediate long-term engraftment of NOD/SCID mice by HIV vectors. Science 283: 682–686

Morgan RA, Anderson WF (1993) Human gene therapy. Annu Rev Biochem 62: 191–217

Moritz T, Dutt P, Xiao X, Carstanjen D, Vik T, Hanenberg H, Williams DA (1996) Fibronectin improves transduction of reconstituting hematopoietic stem cells by retroviral vectors: evidence of direct viral binding to chymotryptic carboxy-terminal fragments. Blood 88: 855–862

Morral N, O'Neal W, Rice K, Leland M, Kaplan J, Piedra PA, Zhou H, Parks RJ, Velji R, Aguilar-Cordova E, Wadsworth S, Graham F L, Kochanek S, Carey KD, Beaudet AL (1999) Administration of helper-dependent adenoviral vectors and sequential delivery of different vector serotype for long-term liver-directed gene transfer in baboons. Proc Natl Acad Sci USA 96: 12816–12821

Mulligan RC (1993) The basic science of gene therapy. Science 260: 926–932

Nakai H, Storm TA Storm TA, Kay MA (2000) Increasing the size of rAAV-mediated expression cassettes in vivo by intermolecular joining of two complementary vectors. Nature Biotechnol 18: 527–532

Naldini L, Verma IM (1999) Lentiviral vectors. In: Friedmann T (ed)The development of human gene therapy. New York, Cold Spring Harbor Laboratory Press, pp. 47–60

Naldini L, Blomer U, Gallay P, Ory D, Mulligan R, Gage FH, Verma IM, Trono D (1996) In vivo gene delivery and stable transduction of nondividing cells by a lentiviral vector. Science 272: 263–267

Nolta JA, Smogorzewska EM, Kohn DB (1995) Analysis of optimal conditions for retroviral-mediated transduction of primitive human hematopoietic cells. Blood 86: 101–110

Olsen JC (1998) Gene transfer vectors derived from equine infectious anemia virus. Gene Ther 5: 1481–1487

Palmer TD, Rosman GJ, Osborne WR, Miller AD (1991) Genetically modified skin fibroblasts persist long after transplantation but gradually inactivate introduced genes. Proc Natl Acad Sci USA 88: 1330–1334

Parks RJ, Chen L, Anton M, Sankar U, Rudnicki MA, Graham FL (1996) A helper-dependent adenovirus vector system: removal of helper virus by Cre-mediated excision of the viral packaging signal. Proc Natl Acad Sci USA 93: 13565–13570

Pfeifer A, Brandon EP, Kootstra N, Gage FH, Verma IM (2001a) Delivery of the Cre recombinase by a self-deleting lentiviral vector: Efficient gene targeting in vivo. Proc Natl Acad Sci USA 10: 1073–1078

Pfeifer A, Kessler T, Yang M, Baranov E, Kootstra N, Cheresh DA, Hoffman RM, Verma IM (2001b) Transduction of liver cells by lentiviral vectors: Analysis in living animals by fluorescence imaging. Mol Ther: 319–322

Poeschla E, Corbeau P, Wong-Staal F (1996) Development of HIV vectors for anti-HIV gene therapy. Proc Natl Acad Sci USA 93: 11395–11399

Poeschla E, Gilbert J, Li X, Huang S, Ho A, Wong-Staal F (1998) Identification of a human immunodeficiency virus type 2 (HIV-2) encapsidation determinant and transduction of nondividing human cells by HIV-2-based lentivirus vectors. J Virol 72: 6527–6536

Poeschla EM, Wong-Staal F, Looney DJ (1998) Efficient transduction of nondividing human cells by feline immunodeficiency virus lentiviral vectors. Nature Med 4: 354–357

Reiser J, Harmison G, (1996) Transduction of nondividing cells using pseudotyped defective high-titer HIV type 1 particles. Proc Natl Acad Sci USA 93: 15266–15271

Shenk T (1996) Adenoviridae: The viruses and their replication. In: Fields BN, Knipe DM, Howley PM (eds) Fields virology. Vol. 2.Philadelphia, Lippincott-Raven Publishers, pp. 2111–2148

St. Louis D, Verma IM (1988) An alternative approach to somatic cell gene therapy. Proc Natl Acad Sci USA 85: 3150–3154

Takahashi M, Miyoshi H, Verma IM, Gage FH (1999) Rescue from photoreceptor degeneration in the rd mouse by human immunodeficiency virus vector-mediated gene transfer. J Virol 73: 7812–7816

Templeton NS, Lasic D (eds) (2000) Gene therapy: therapeutic mechanisms and strategies. New York, Marcel Dekker, Inc.

Tisdale JF, Hanazono Y, Sellers SE, Agricola BA, Metzger ME, Donahue RE, Dunbar CE (1998) Ex vivo expansion of genetically marked rhesus peripheral blood progenitor cells results in diminished long-term repopulating ability. Blood 92: 1131–1141

Trono D (2000) Lentiviral vectors: turning a deadly foe into a therapeutic agent. Gene Ther 7: 20–23

Verma IM (1990) Gene therapy. Sci Am 263: 68–72, 81–84

Weinberg JB, Matthews TJ, Cullen BR, Malim MH (1991) Productive human immunodeficiency virus type 1 (HIV-1) infection of nonproliferating human monocytes. J Exp Med 174: 1477–1482

Xu K, Ma H, McCown TJ, Verma IM, Kafri T (2001) Generation of a stable cell line producing high-titer self-inactivating lentiviral vectors. Mol Ther 3: 97–104

Yan Z, Zhang Y, Duan D, Engelhardt JF (2000) From the cover: trans-splicing vectors expand the utility of adeno-associated virus for gene therapy. Proc Natl Acad Sci USA 97: 6716–6721

Yeh P, Perricaudet M (1997) Advances in adenoviral vectors: from genetic engineering to their biology Faseb J 11: 615–623

Yu H, Rabson AB, Kaul M, Ron Y, Dougherty JP (1996) Inducible human immunodeficiency virus type 1 packaging cell lines. J Virol 70: 4530–4537

Yu SF, von Ruden T, Kantoff PW, Garber C, Seiberg M, Ruther U, Anderson WF, Wagner EF, Gilboa E (1986) Self-inactivating retroviral vectors designed for transfer of whole genes into mammalian cells. Proc Natl Acad Sci U S A 83: 3194–3198

Zufferey R, Donello JE, Trono D, Hope TJ, (1999) Woodchuck hepatitis virus posttranscriptional regulatory element enhances expression of transgenes delivered by retroviral vectors. J Virol 73: 2886–2892

# Gene Therapy in the Central Nervous System

*M. Barkats[1], A. Bemelmans[1], S. Brun[1], O. Corti[1], C. Sarkis[1], and J. Mallet[1]*

## Introduction

Neurodegenerative diseases represent a substantial burden for society in terms of both patient suffering and socio-economic costs. For example, in Europe nearly 3 million persons suffer from stroke, over 3.5 million from dementia (with a higher proportion of women than men) and about 1 million from Parkinson's disease (PD). Overall, diseases of the nervous system, including eye diseases, account for approximately 25% of all health costs in the EU, which in money terms translates into hundreds of billions of Euros. With increasing life expectancy, this cost is likely to greatly increase, since the prevalence of dementia is 18% of the population between 80 and 85 years of age and 32% between 85 and 90. Similarly, PD affects more than 10% of the population after the age of 85. There is no cure for most of the neurodegenerative diseases, and the very few available treatments are unsatisfactory. The urgent need to develop therapies will rely on the unravelling of the pathophysiological mechanisms underlying these afflictions and on the development and use of post-genomics and cell/gene transfer technologies.

The recent and considerable developments in behavioural, imaging and electrophysiological methods will lead to the identification of the brain and spinal cord networks and neural mechanisms underlying the generation of symptoms associated with neurodegenerative disorders. These investigations are crucial in the selection of the brain regions and cells that are to be investigated at the molecular level. Considerable efforts are being devoted to the deciphering in molecular terms of the mechanisms underlying cell death in cellular model systems and in the above disease states. Such an approach is the most logical one to uncover novel and rational targets and treatments for these diseases with an integrated and convergent approach based on functional genomics (transcriptomics, proteomics in a broad sense) and genetics.

Some of the above diseases have purely genetic forms. They are thus amenable to positional cloning approaches that do not rely on any hypothesis about their pathophysiological mechanisms. These approaches have provided important clues for amyotrophic lateral sclerosis (ALS), Huntington's disease (HD) early onset Alzheimer's disease (AD) and PD, although much further work is still

[1] Laboratoire de Génétique Moléculaire de la Neurotransmission et des Processus Neurodégénératifs, Centre National de la Recherche Scientifique UMR 7091, Bât CERVI, Hôpital Pitié Salpétriére, 75013 Paris, France

Mallet/Christen
Neurosciencess at the Postgenomic Era
© Springer-Verlag Berlin Heidelberg 2003

needed to unravel in molecular terms the processes that underlie these diseases. The genetics of late onset PD and AD are complex and most probably heterogeneous and polygenic in nature. The genetic factors need to be teased apart and will provide further entry points for understanding cell death mechanisms. Such studies of complex diseases are opportune, following the recent development of the human genome projects. Particularly, the availability of very dense, single-nucleotide polymorphism (SNP) maps makes it possible to envisage association studies across the whole genome in the near future. This should greatly expedite the identification of susceptibility genes.

Once a gene or a protein has been implicated in the cellular damage or cell death process, it is crucial to identify all protein partners and members of corresponding functional networks at the cellular and system levels. Such members will provide potential novel drug targets. Cluster analysis will be instrumental in identifying networks. This process is still in its early stages and will require the involvement of computer scientists and mathematicians to develop the new algorithm methods that will be required.

Independently of such genetic studies, recent advances in our understanding of the pathophysiological mechanisms underlying these diseases have led to the identification of several endogenous factors that could be of therapeutic value. These are trophic factors such as nerve growth factor (NGF), brain-derived neurotrophic factor (BDNF), glial cell line-derived neurotrophic factor (GDNF), neurotrophin-3 (NT3), neurotrophin-4/5 (NT4/5), enzymes involved in the clearing of free radicals, such as the superoxide dismutase (SOD), and anti-apoptotic proteins. These factors have been shown to efficiently protect neurons from decay or death and could lead to protective therapies. When a neurodegenerative disease has progressed for a long time, large numbers of neurons have already degenerated and such neuroprotective strategies are no longer effective. An alternative approach, which seems efficient in some situations, is to locally restore neurotransmitter levels that are depleted following the death of the corresponding neurons. Such strategies are called restorative therapies.

This review will focus on recent developments in the field of gene therapy for neurological diseases, with particular attention to the work that is carried out in our laboratory. Proofs of concept that have been obtained will be presented together with the investigations that remain to be pursued to bring this field to practical fruition.

## Advantages of Gene Therapy for Treating Neurological Diseases

The discoveries of factors of potential therapeutic value to treat neurological diseases have, to date, not been translated into clinical practice because these factors are proteins that cannot be delivered as such to the desired region of the nervous system. Systemic (intravenous) injection of the therapeutic protein to the patient is problematic because of side effects such as weight loss, inflammation and fever. Since such proteins are rapidly degraded in the bloodstream, the therapeutic dose would be above the tolerable threshold and could trigger side effects, like those that led to the cessation of clinical trials based on CNTF injection to ALS patients.

One possible way to overcome these limitations is to deliver the protein factor directly to the site in the nervous system where it is expected to exert its therapeutic effect. Such sites are, for example, the striatum in the case of PD and HD, the spinal cord in the case of ALS, and the retina in the case of retinal degenerative disorders. Under some circumstances, this strategy has  proven useful for testing the acute effect of particular factors in animals. However, local protein injections cannot be contemplated for the sustained delivery of the therapeutic agent whose effect is required for months or even years in humans. It may be that such hurdles will be overcome by gene therapy approaches, in which genes are delivered locally to the nervous system. The genes of interest, driven by appropriate promoter sequences, will use the endogenous machinery of host cells to ensure sustained production of the therapeutic factors. In this instance, the drug is the DNA fragment that is being delivered to the nervous system.

A major breakthrough in the field occurred following the discovery in our laboratory that a non-neurotropic virus, the adenovirus, can efficiently transduce both neurons and glial cells (Le Gal La Salle et al. 1993). This finding opened a new field of endeavour and greatly stimulated the development of gene transfer protocols in the nervous system, both for the study of biological mechanisms and for clinical purposes. Recently, other vectors, such as those derived from adeno-associated viruses (AAV) or lentiviruses, were also found to be very effective for transducing nerve cells.

## Parkinson's Disease as a Prototype for CNS Gene Therapy

PD is a neurological disorder that involves the progressive loss of dopaminergic neurons from the substantia nigra, a brain structure that innervates the striatum, leading to specific motor impairment (tremor, rigidity and akinesia) that is sometimes associated with cognitive impairment (Poewe and Wenning, 1996).

Classical oral administration of L-Dopa (a substance that crosses the blood-brain barrier and is thereafter transformed into the neurotransmitter called dopamine) can significantly improve motor function during the first years of treatment. But, as the neurodegeneration progresses, this therapy becomes less effective, requiring higher doses of L-Dopa, which leads to deleterious side effects such as dyskinesia. L-Dopa or other related drugs can prevent the appearance of symptoms (tremor, rigidity), but unfortunately, these drugs do not treat the cause and the disease progresses until death.

A promising therapeutic strategy has recently been described that is based on grafts of dopamine-secreting cells. The efficacy of intrastriatal grafts of mesencephalic tissue, which are very rich in dopaminergic cells was first shown in well-characterized animal models from rats and monkeys. In humans, such grafts have yielded an amelioration of symptoms for more than three years (Hauser et al. 1999). This strategy is interesting, but its general use is hampered by the need for tissue from several human foetuses to treat each patient, a requirement that raises considerable safety, regulatory and ethical issues.

A neurosurgical method (subthalamic electrostimulation) has also been proposed (Benabid et al. 1998; Limousin et al., 1998). Electrodes are implanted in deep brain regions and electrically stimulate subthalamic nuclei. This treatment allows a reduction of up to 50% of the drugs needed for PD treatment and noticeably decreases dyskinesia. The electrostimulating system has to remain within the brain (electrode) and in the thorax (battery). Despite its efficiency, this method has drawbacks, such as providing a niche for bacteria, becoming "untuned" by microwaves or magnetic systems, and requiring the battery charging . Obviously, a gene therapy treatment that could be done once and work for a long period of time would be a major leap forward.

Gene therapy studies for PD were stimulated by the availability of animal models in rats and primates. They involved both the restorative and protective strategies. We initially devised a restorative approach based on an adenoviral vector expressing the human tyrosine hydroxylase isoform1(hTH1), which was introduced into the brain of rats impaired with drug-induced (6-OHDA) Parkinson's symptoms (Horellou et al. 1994). This treatment induced a 25% improvement in functional tests. This work was followed by many other studies using various vectors. Most notably, Fan et al. (1998) induced full recovery in 6-OHDA-lesioned rats after injection of two AAV vectors bearing eitherthe hTH gene or the AADC gene.

The protective strategy offers a better prospect for the long-term treatment of the disease. Initially, our laboratory used an adenovirus expressing the trophic factor GDNF (Ad-GDNF), which was isolated recently and is structurally related to members of the transforming growth factor superfamily. This factor was first identified by its ability to promote the survival and neurite outgrowth of embryonic dopaminergic (DA) neurons in vitro (Lin et al. 1993). Intracerebral administration of GDNF over the substantia nigra (SN) was then shown to completely prevent nigral cell death and atrophy in a rat model of PD (Sauer et al. 1995).

We tested the effect of Ad-GDNF in vivo using a model of progressive DA cell degeneration resulting from the unilateral injection of 6-OHDA into the striatum. When Ad-GDNF was injected into the same site six days before the 6-OHDA lesion, the production of GDNF was observed at both DA nerve terminal and nigral cell bodies following retrograde transport of the virus. Three weeks after the lesion, there were twice as many TH-immunoreactive cell bodies in the SN of the Ad-GDNF-treated animals as in control animals. Furthermore, the amphetamine-induced behavioural asymmetry (a behavioural marker for DA depletion) was markedly reduced in the Ad-GDNF-treated rats (Bilang-Bleuel et al. 1997). This study provided the first proof-of-principle of the therapeutic potential of the protective strategy for PD and neurodegenerative diseases in general.

A similar neuroprotective effect of an adenovirus encoding GDNF, which was reported by Choi-Lundberg and collaborators (1997) in the 6-OHDA progressive lesion model of PD after direct injection of Ad-GDNF near the SN, confirmed our initial observation on dopaminergic cell body rescue. However, in this case no striatal reinnervation was observed.

More recently these studies have been extended to nonhuman primate models of PD. Kordower et al. (2000) injected a lentiviral vector expressing GDNF (lenti-

GDNF) into the striatum and substantia nigra of nonlesioned, aged rhesus monkeys or young adult rhesus monkeys treated one week earlier with 1-methyl-4-phenyl-1,2,3,6-tetrahydropyridine (MPTP). Extensive GDNF expression with anterograde and retrograde transport was seen in all animals. In aged monkeys, lenti-GDNF augmented dopaminergic function. In MPTP-treated monkeys, lenti-GDNF reversed functional deficits and completely prevented nigrostriatal degeneration. Additionally, lenti-GDNF injections to intact rhesus monkeys revealed long-term gene expression (eight months). In MPTP-treated monkeys, lenti-GDNF treatment reversed motor deficits in a hand-reach task (Kordower et al. 2000). These data further indicate that GDNF gene transfer can prevent nigrostriatal degeneration and might be a viable therapeutic strategy for PD patients.

## Motoneuron Diseases and Nerve Injuries.

### Retrograde Transport of Adenoviral Vectors

The property of adenoviral vectors to be retrogradely transported provides a convenient access to motoneurons and should allow nerve re-growth and repair. We have shown that specific gene transfer into motor neurons can be obtained by peripheral intramuscular injections of recombinant adenoviruses (Finiels et al. 1995). These vectors are retrogradely transported from muscular endplates to motor neuron cell bodies. Gene transfer can thus be specifically targeted to each level of the spinal cord by appropriate choice of the injected muscle. This new therapeutic protocol allows specific targeting of motor neurons without lesioning the spinal cord, and should avoid undesirable side effects associated with systemic administration of therapeutic factors. Most interestingly, the retrograde transport also occurs along lesioned nerves, thereby allowing the easy delivery of factors that promote nerve outgrowth.

### Neuroprotection of Facial Motoneurons

The retrograde transport of adenovirus expressing growth factors was used to confer neuroprotection to facial motoneurons. Giménez y Ribotta and al.(1997) studied the protective effect of injecting AD-GDNF or BDNF into the nasolabial and lower lip muscles of newborn rats following the death of the axotomized facial neurons. A significantly better survival of axotomized motor neurons was recorded one week after surgery in rats pretreated with Ad-BDNF (34.5%) or Ad-GDNF (41.2%) than in rats pretreated with the control Ad-βgal (Gimenez y Ribotta et al., 1997). Similar results were reported by Gravel et al.(1997) using adenoviruses encoding CNTF or BDNF. The same procedure of intramuscular injection of Ad-GDNF before axotomy was also used to confer neuroprotection to lumbar spinal motoneurons (Baumgartner and Shine 1997) .

## Increased Transgene Expression in Motoneurons of ALS Mice and in Botulinum-Injected Mice

These findings provided evidence for the potential of recombinant adenoviruses for the efficient long-term protection of motor neurons. They opened the way for a possible treatment of severe degenerative diseases such as ALS for which there is no classical treatment with proven efficiency. ALS, also called "Lou Gehrig's disease", is an adult-onset, progressive neurodegenerative disease that affects cortical, spinal and bulbar motoneurons and manifests itself by muscular weakness, atrophy and spasticity. Progression is relatively rapid and leads to paralysis and death, usually within two to five years (Wong et al. 1998). Most ALS cases are sporadic, with about 10% autosomal dominant inherited cases; the two forms of the disease are clinically indistinguishable.

A large number of point mutations have been discovered in the copper/zinc-superoxide dismutase (SOD1) gene in about 20% of familial ALS (FALS) cases (Rosen et al. 1993; Deng et al. 1993). SOD1 is a cytosolic enzyme that plays a critical role in preventing oxidative stress-induced cell damage by scavenging the superoxide radical and converting it into hydrogen peroxide and oxygen (Gutteridge and Halliwell 2000). Although the mechanism by which the SOD1 mutations are selectively toxic to motoneurons is unknown, some of the mutations cause motoneuron disease when expressed in transgenic mice (Bruijn et al. 1997; Gurney 1994; Ripps et al. 1995; Wong et al. 1995). For example, transgenic mice that overexpress the G93A mutation of the human SOD1 (glycine to alanine substitution at position 93) develop a motoneuronal disorder with a progressive death of motoneurons and vacuolar degeneration of mitochondria, leading to paralysis and death at about five months of age (Dal Canto and Gurney 1995; Gurney 1994).

The value of the gene delivery method in motoneurons using retrograde axonal transport of adenoviruses in ALS remained unclear in view of the numerous studies that reported denervation and impairment of axonal transport in the mouse models of the disease (Tu et al. 1996; Warita et al. 1999; Williamson and Cleveland 1999; Zhang et al. 1997). Using adenoviruses encoding reporter genes as retrograde tracers, we assessed the capacity of motoneurons to take up and retrogradely transport adenoviral particles injected into the muscles of transgenic mice expressing the G93A SOD1 mutation. Surprisingly, transgene expression in the motoneurons was significantly higher in symptomatic mice than in control or presymptomatic mice (Millecamps et al. 2001), and it was not a consequence of the wtSOD overproduction. The unexpectedly high level of motoneuron retrograde transduction resulted, at least in part, from the axonal sprouting associated with the disease. Indeed, treatment with botulinum toxin, which is known to induce nerve sprouting at neuromuscular junctions, dramatically enhanced gene transfer to motoneurons innervating the injected muscles both in wild-type and ALS mice (Millecamps et al. 2002). These findings have major applications for gene delivery of exogenous factors to motoneurons using axonal retrograde transport of viral vectors and demonstrate the rationale for using intramuscular injections of adenoviruses to overexpress proteins of interest in motor neuron diseases.

## Requirements for Gene Delivery to Become an Effective Therapy

As discussed above several factors have already been found to exhibit therapeutic effects in animal models for several neurodegenerative diseases. Without any doubt, additional potential therapeutic factors will be identified in the near future, given the current flurry of investigations into the pathogenic mechanisms underlying these diseases. However, several crucial issues need to be addressed to make these factors effective therapeutic agents.

### Safety of the Vector

The first issue concerns the possible toxicity of the viruses: they may interfere with the endogenous cellular machinery and ultimately cause the death of the transduced cells. However, the effect is dose-dependent and in most instances no cytopathic effects have been observed with the doses of viruses needed to obtain a therapeutic effect.

A more important issue is that the presence of the vector in the body may trigger an immune response to any of three different classes of molecules: 1) the therapeutic gene product, 2) the viral proteins encoded by genes carried on the viral backbone, and 3) the viral capsid proteins. The expression of the therapeutic factors is in most cases not problematic, since most of them are proteins that are normally present in nervous tissues during development and/or in the adult. Thus, they are not recognized as foreign proteins and do not generate an immune response. The second class is of greater concern, with vectors such as the first generation adenoviral or herpes vectors, which still express endogenous viral genes. These gene products can trigger an immune response that may ultimately lead to the degeneration of transduced cells (Wood et al. 1996; Yang et al. 1994). This is, however, not an issue for gutless adenoviral vectors, the AAV and the lentivirus-derived vectors; all of the viral genes have been deleted from these vectors. The last class includes the capsid proteins of the virus that will only be present for a few days after injection. There are, however, ways to circumvent the immune response raised by these short-lasting exogenous proteins. One may, for example, temporarily blind the immune response during vector administration.

### Viral Vectors for CNS Applications

No single vector is, at present, the best for treating all neurodegenerative diseases or even for carrying the various therapeutic genes that might be considered for treating a single disease. It is likely that several vectors, including those derived from adenovirus, AAV and HIV, will be important in the gene therapy of the CNS.

Adenoviruses have several features that make them attractive candidates for use as gene therapy vectors. These non-enveloped viruses, which contain a well-characterized 36 kb double-strand DNA genome, can be easily grown and purified in large quantities, are able to transduce a wide variety of dividing and postmitotic cells, and have a broad natural tropism that can be altered by modifications of the capsid (Krasnykh et al. 1996; Wickham et al. 1997). However, first-generation adenoviral vectors (i.e., vectors deleted for E1), although permitting long-term expression in rat brain, induce a cellular immune response toward transduced cells that would be deleterious in a human setting. Low-level expression of viral genes is, to a large extent, responsible for direct vector toxicity, inflammation in the transduced tissue, and the strong cellular immune response against the virus. Numerous efforts have been undertaken to minimize these adverse effects and to develop safer and more efficient adenovirus vectors. A new approach in the design of adenoviral vectors with improved capacity and safety is the development of new, fourth-generation adenoviral vectors that lack all viral coding sequences (Mitani et al. 1995). These adenoviral vectors, also designated as "gutless'" vectors or helper-dependent adenoviral vectors, prevent low-level expression of viral genes that leads to toxicity and immune-mediated removal of transduced cells. Furthermore, they can accommodate large regulatory DNA regions, in contrast to first-generation adenoviral vectors, which can incorporate only 8 kb of insert DNA. "Gutless" vectors were developed to address some of the limitations posed by the other generations of adenoviral vectors while retaining their advantages. Finally, adenoviral vectors can be retrogradely transported through motor neuron axons after a peripheral intramuscular injection, in contrast to lentiviral vectors.

Lentiviral vectors integrate into the genome of the transduced cells and therefore persist and allow the expression of the transgene for a very long time. A single injection may suffice for very long-term expression in man. As they do not express or encode viral genes, the transduced cells are not recognized by the host immune system. Moreover, lentiviral vectors display several advantages. Contrary to the widely used oncoretroviral vectors, lentiviral vectors are able to transduce post-mitotic cells, including neurons, with high efficiency (Naldini et al. 1996), and their tropism can be modified and easily broadened by using heterologous envelop proteins. It is thus possible to obtain lentiviral vectors targeting astrocytes with high specificity (unpublished data) or targeting the retinal pigment epithelium cells of the eye (Auricchio et al. 2001). We are currently working with lentiviral vectors containing a DNA flap sequence, which increases the transduction efficiency of the vector (Zennou et al. 2000, 2001).

The efficacy of both adenoviral and lentiviral vectors has been demonstrated in various therapeutic models for which each vector displays a specific advantage. Nevertheless, to our knowledge, there is currently no satisfactory, large industrial-scale production available for any viral vector. In this respect, the fact that only small amounts of a vector are necessary in most CNS applications represents a real advantage.

### Control of Transgene Expression

The issue of the control of transgene expression constitutes an important chapter in the field of gene therapy. The various aspects to be addressed include the longevity of the expression of the transgene and its strength, tissue and cell specificity as well as its control by external agents.

The regulatory sequences that control the expression of the exogenous genes often do not remain active, thereby altering the longevity of the effects of the treatments. For example, the gradual inactivation of the strong promoter elements from the RSV or Cytomegalovirus (CMV) that are generally used may explain the decline in expression over time following infection (Palmer et al. 1991). This issue is all the more important because, in the case of neurodegenerative diseases, expression of the therapeutic gene is required for several years. Results in our laboratory indicate that, in an adenoviral backbone, the use of cellular promoters greatly increases the longevity of transgene expression as compared to viral promoters (Navarro et al. 1999). This, together with the use of "gutless" viruses, should allow the design of vectors that fulfill the above requirement. In the case of lentiviral vectors, the vector genome integrates into the host cell. Thus, if such vectors are used with an adequate promoter, the transgene expression might continue as long as the transduced cell survives. An expression lasting more than one year has already been described for several cell types of the rodent.

Cellular promoters have the added benefit of allowing precise targeting of expression to particular cell types. We injected into the rat hippocampus an adenoviral vector expressing the *E. coli* LacZ reporter gene under the control of the rat neuron-specific enolase (NSE) promoter. The pattern of X-gal staining showed that neurons were preferentially transduced and β-galactosidase expression persisted for six months (Navarro et al. 1999). In another study we generated adenoviral constructs carrying a neuron-restrictive silencer element (NRSE) upstream from a ubiquitous promoter to target luciferase gene expression to neurons by repressing ectopic expression in non-neuronal cells. After intramuscular injection of adenoviral constructs containing 6 or 12 NRSE sequences, strong luciferase expression was obtained in neuronal cells after retrograde transport of the adenovirus whereas almost no expression was observed in non-neuronal muscle cells (Millecamps et al. 1999). A report by Morelli et al. (1999) also discussed the use of neuronal and glial-specific promoters to target the expression of a transgene to predetermined brain cells.

Another means to restrict the expression of a transgene is to target the delivery of vectors to specific cell types. Changes in the patterns of tropism as well as transduction efficiency can be obtained by pseudotyping in the case of vectors such as lentiviral vectors. We have constructed and tested three different envelope proteins (or G proteins) from rhabdovirus: the G protein of 1) the vesicular stomatitis virus (VSV, vesicular genus), 2) the rabies virus (PV strain, lyssavirus genus) and 3) the mokola virus (MOK, lyssavirus genus). VSV-G was at least 10 times more abundant than MOK-G or PV-G at the virion surface, as assessed by quantitative Western blotting. The efficiency of the three pseudotypes was tested in vitro using seven different neural cell lines and rat cortical primary neurons. VSV- and MOK-pseudotyped vectors had similar efficiencies of transduction, higher than those of PV-pseudotyped vectors. The cellular tropism was investigated following intracerebral injection of the vectors. In the case of VSV-pseudotyped vectors, the transduced cells included both astrocytes and neurons, with a preferential tropism toward astrocytes when the CMV promoter was used and toward neurons with the phosphoglycerate kinase (PGK) promoter. MOK and PV vectors displayed a specific tropism for astrocytes with both the CMV and PGK promoters (Sarkis et al., unpublished). These vectors therefore could be convenient tools available for targeting astrocytes in vivo if these first results are confirmed. This feature may be quite useful for gene therapy protocols that involve secreted therapeutic factors, a situation that is likely to prevail in the case of neurodegenerative diseases. Then, astrocytes will efficiently produce and secrete these factors that will protect neurons whose integrity and metabolism will not be affected by the presence of the vector.

An important consideration in gene therapy is also to minimise the amount of vector to be used in order to reduce its potential toxicity. One way to reduce the viral doses while maintaining high levels of expression is to improve the expression per molecule of vector. Much of the effort aimed at enhancing transgene expression has been directed at boosting transcription. We have recently investigated the enhancement of transgene expression in neuronal and glial phenotypes with post-trancriptional regulatory elements. In particular we tested, alone and in combination, the effects of 1) the well-known regulatory element present in the woodchuck hepatitis virus (WPRE; Loeb et al. 1999), 2) a fragment of the rat tyrosine hydroxylase (TH) mRNA that confers increased stability to mRNA in a rat neuronal cell line (Paulding and Czyzyk-Krzeska 1999), 3) a fragment of the 3'UTR of rat tau mRNA (Aronov et al. 1999), and 4) a fragment of the 5' UTR of the human amyloid precursor protein (APP) mRNA that increases translation in human neuronal and glial cell lines (Rogers et al.,1999). These elements were introduced in plasmid and lentiviral vectors coding for luciferase. These constructs allow high-level gene expression in both neuronal and glial cells. The highest level of expression was obtained with WPRE, which enhanced the luciferase expression about four-fold. Most importantly, further enhancements were obtained when WPRE was combined with the other aforementioned sequences. Particularly, the combination of WPRE and both tau 3'UTR and APP 5'UTR increased the basal level of expression up to 26-fold in both the plasmid and lentiviral contexts (Brun et al., submitted for publication).

Finally, it is important to stress that, for clinical application, a vector should allow adaptation of the treatment to the needs of the patient and termination of the therapy if necessary. For example, in the case of neurodegenerative diseases, the local and regulated production of trophic factors is crucial since excessive expression may be deleterious to some of the cells. The development of efficient gene regulatory systems based on specific transcription factors that respond to exogenous drugs has allowed the integration of inducible expression cassettes into gene delivery vectors. One particularly thoroughly studied circuit of transcriptional activation uses elements of the tetracycline-resistance operon *of Escherichia coli* transposon 10 to create two novel transactivators that are controlled either negatively (tet-off system) or positively (tet-on system) by tetracycline (Gossen and Bujard 1992; Gossen et al. 1995). The efficacy of these transactivators for the regulation of foreign gene expression has been demonstrated in a variety of mammalian organs. We and other laboratories have investigated the potential of these systems for gene therapy in the CNS. As was the case in several other studies, efficient regulation was only possible when the ratio of the transactivator virus to reporter virus was adapted. We have simplified the original two-vector-based tet-off system for future clinical application by designing  a single adenovirus for tet-regulated expression of therapeutic genes in the CNS (AdGPK-tet-hTH-1;Corti et al. 1996). This system allowed tet-controlled synthesis of human TH in cultured nerve cells (Corti et al. 1999a). Moreover, efficient and reversible control of TH expression was obtained in a rat model of PD, both by an ex vivo gene transfer approach based on grafting of genetically modified human neural progenitors and by direct intracerebral injection of regulatable adenovirus (Corti et al. 1999b).

## Conclusions

With the recent advances in  behavioural, imaging  and electrophysiological methods, coupled with genetics and functional genomics approaches, there is much hope that novel and rational treatments for curing neurodegenerative diseases will become available in the near future. Gene therapy protocols are likely to occupy important places  among the potential treatments. Proof of concept has recently been established that various vectors can appropriately produce and deliver, in the nervous system, therapeutic factors such as trophic factors that protect neurons from degenerating. Important issues such as the safety of the vectors and the longevity and regulation of the expression of the transgenes are now receiving much attention in order to make gene therapy approaches effective. With these new developments gene therapy should soon play an important role in the clinical practice of neurology.

## Acknowledgments

The authors wish to thank Sue Orsoni for reading  the manuscript. This work was supported by the Centre National de la Recherche  Scientifique, the Association Française contre les Myopathies, the Institut pour la Recherche  sur la Moelle Epinière  and  Retina France.

## References

Aronov S, Marx R,  Ginzburg I (1999) Identification of 3'UTR region implicated in tau mRNA stabilization in neuronal cells. J Mol Neurosci 12:131–145.

Auricchio A, Kobinger  G, Anand V, Hildinger M, O'Connor E, Maguire AM, Wilson JM, Bennett J (2001) Exchange of surface proteins impacts on viral vector cellular specificity and transduction characteristics: the retina as a model. Human Mol Genet 10 ; 3075–3081.

Baumgartner BJ, Shine HD (1997) Targeted transduction of CNS neurons with adenoviral vectors carrying neurotrophic factor genes confers neuroprotection that exceeds the transduced population. J Neurosci 17: 6504–6511.

Benabid AL, Benazzouz A, Hoffmann D, Limousin P, Krack P, Pollak P (1998) Long-term electrical inhibition of deep brain targets in movement disorders. Movement Disord 13: 119–125.

Bilang-Bleuel A, Revah F, Colin P, Locquet I, Robert JJ, Mallet J, Horellou P (1997) Intrastriatal injection of an adenoviral vector expressing glial-cell- line-derived neurotrophic factor prevents dopaminergic neuron degeneration and behavioral impairment in a rat model of Parkinson disease. Proc Natl Acad Sci USA 94: 8818–8823.

Bruijn LI, Becher MW, Lee MK., Anderson KL, Jenkins NA, Copeland NG, Sisodia, SS, Rothstein JD, Borchelt DR, Price DL, Cleveland DW (1997) ALS-linked SOD1 mutant G85R mediates damage to astrocytes and promotes rapidly progressive disease with SOD1-containing inclusions. Neuron 18,; 327–338.

Choi-Lundberg DL, Lin Q, Chang YN, Chiang YL., Hay CM, Mohajeri H, Davidson BL, Bohn MC (1997) Dopaminergic neurons protected from degeneration by GDNF gene therapy. Science 275: 838–841.

Corti O, Horellou P, Colin P, Cattaneo E., Mallet J (1996) Intracerebral tetracycline-dependent regulation of gene expression in grafts of neural precursors. Neuroreport 7: 1655–1659.

Corti O., Sabate O, Horellou P, Colin P, Dumas S, Buchet D, Buc-Caron MH, Mallet J (1999a) A single adenovirus vector mediates doxycycline-controlled expression of tyrosine hydroxylase in brain grafts of human neural progenitors. Nature Biotechnol 17: 349–354.

Corti O, Sanchez-Capelo A, Colin P, Hanoun N, Hamon M, Mallet J (1999b). Long-term doxycycline-controlled expression of human tyrosine hydroxylase after direct adenovirus-mediated gene transfer to a rat model of Parkinson's disease. Proc Natl Acad Sci USA 96: 12120–12125.

Dal Canto MC, Gurney,ME (1995) Neuropathological changes in two lines of mice carrying a transgene for mutant human Cu,Zn SOD, and in mice overexpressing wild type human SOD: a model of familial amyotrophic lateral sclerosis (FALS). Brain Res 676,:25–40.

Deng HX, Hentati A, Tainer JA, Iqbal Z, Cayabyab A, Hung WY, Getzoff ED, Hu P, Herzfeldt, B., Roos, R. P., and et al. (1993). Amyotrophic lateral sclerosis and structural defects in Cu,Zn superoxide dismutase. Science 261, 1047–51.

Fan DS, Ogawa M, Fujimoto KI, Ikeguchi K, Ogasawara Y, Urabe M, Nishizawa M, Nakano I, Yoshida M, Nagatsu I, Ichinose H, Nagatsu T, Kurtzman GJ, Ozawa K (1998) Behavioral recovery in 6-hydroxydopamine-lesioned rats by cotransduction of striatum with tyrosine hydroxylase and aromatic L- amino acid decarboxylase genes using two separate adeno-associated virus vectors. Human Gene Ther 9: 2527–2535.

Finiels F, Gimenez y Ribotta M, Barkats,M, Samolyk ML, Robert J, Privat A, Revah F, Mallet J (1995) Specific and efficient gene transfer strategy offers new potentialities for the treatment of motor neurone diseases. Neuroreport 6: 2473–2478.

Gimenez y Ribotta M, Revah F, Pradier L, Loquet I, Mallet, J, Privat A (1997) Prevention of motoneuron death by adenovirus-mediated neurotrophic factors. J Neurosci Res 48: 281–285.

Gossen M, Bujard H (1992) Tight control of gene expression in mammalian cells by tetracycline-responsive promoters. Proc Natl Acad Sci USA 89: 5547–5551.

Gossen M, Freundlieb S, Bender G, Muller G., Hillen W, Bujard H (1995) Transcriptional activation by tetracyclines in mammalian cells. Science 268: 1766–1769.

Gravel C, Gotz R, Lorrain A, Sendtner M (1997) Adenoviral gene transfer of ciliary neurotrophic factor and brain- derived neurotrophic factor leads to long-term survival of axotomized motor neurons. Nature Med 3: 765–770.

Gurney ME (1994) Transgenic-mouse model of amyotrophic lateral sclerosis. New Engl J Med 331: 1721–1722.

Gutteridge JM, Halliwell B (2000) Free radicals and antioxidants in the year 2000. A historical look to the future. Ann NY Acad Sci 899,:136–147.

Hauser RA, Freeman TB, Snow BJ, Nauert M, Gauger L, Kordower H, Olanow CW (1999) Long-term evaluation of bilateral fetal nigral transplantation in Parkinson disease. Arch Neurol 56: 179–187.

Horellou P, Vigne E, Castel MN, Barneoud, P, Colin, P, Perricaudet,M, Delaere P, Mallet J (1994) Direct intracerebral gene transfer of an adenoviral vector expressing tyrosine hydroxylase in a rat model of Parkinson's disease. Neuroreport 6: 49–53.

Kay MA, Glorioso JC, Naldini L (2001) Viral vectors for gene therapy: the art of turning infectious agents into vehicles of therapeutics. Nature Med 7: 33–40.

Kordower J.H, Emborg ME, Bloch J, Ma SY, Chu Y, Leventhal L, McBride J, Chen EY, Palfi S, Roitberg BZ, Brown WD, Holden JE, Pyzalski R, Taylor,MD, CarveyP, Ling, Z, Trono D, Hantraye P, Deglon N, Aebischer P (2000) Neurodegeneration prevented by lentiviral vector delivery of GDNF in primate models of Parkinson's disease. Science 290: 767–773.

Krasnykh VN, Mikheeva GV, Douglas JT, Curiel DT (1996) Generation of recombinant adenovirus vectors with modified fibers for altering viral tropism. J Virol 70: 6839–6846.

Le Gal La Salle G, Robert JJ, Berrard S, Ridoux V, Stratford-Perricaudet LD, Perricaudet M, Mallet J (1993) An adenovirus vector for gene transfer into neurons and glia in the brain. Science 259: 988–990.

Limousin P, Krack P, Pollak P, Benazzouz A, Ardouin,C, Hoffmann D, Benabid A L (1998) Electrical stimulation of the subthalamic nucleus in advanced Parkinson's disease. New Engl J Med 339: 1105–1111.

Lin LF, Doherty DH, Lile JD, Bektesh S, Collins F (1993). GDNF: a glial cell line-derived neurotrophic factor for midbrain dopaminergic neurons. Science 260:1130–1132.

Loeb JE, Cordier WS, Harris ME, Weitzman MD, Hope TJ (1999) Enhanced expression of transgenes from adeno-associated virus vectors with the woodchuck hepatitis virus posttranscriptional regulatory element: implications for gene therapy. Human Gene Ther 10,: 2295–2305.

Millecamps S, Kiefer H, Navarro V, Geoffroy MC, Robert JJ, Finiels F, Mallet J, Barkats M (1999) Neuron-restrictive silencer elements mediate neuron specificity of adenoviral gene expression. Nature Biotechnol 17: 865–869.

Millecamps S, Nicolle D, Ceballos-Picot I, Mallet, J, Barkats M (2001) Synaptic sprouting increases the uptake capacities of motoneurons in amyotrophic lateral sclerosis mice. Proc Natl Acad Sci USA 98: 7582–7587.

Millecamps S, Mallet J, Barkats M (2002) Adenoviral retrograde gene transfer in motoneurons is greatly enhanced by prior intramuscular inoculation with botulinum toxin. Human Gene Ther 13: 225–232.

Mitani K, Graham FL, Caskey CT, Kochanek S (1995) Rescue, propagation, and partial purification of a helper virus- dependent adenovirus vector. Proc Natl Acad Sci USA 92: 3854–3858.

Morelli AE, Larregina AT, Smith-Arica J, Dewey RA, Southgate,TD, Ambar B, Fontana A, Castro MG, Lowenstein PR (1999) Neuronal and glial cell type-specific promoters within adenovirus recombinants restrict the expression of the apoptosis-inducing molecule Fas ligand to predetermined brain cell types, and abolish peripheral liver toxicity. J Gen Virol 80: 571–583.

Naldini L, Blomer U, Gallay P, Ory D, Mulligan R, Gage FH, Verma IM, TronoD (1996) In vivo gene delivery and stable transduction of nondividing cells by a lentiviral vector. Science 272: 263–267.

Navarro V, Millecamps S, Geoffroy MC, Robert JJ. Valin A, Mallet J, Gal La SalleG.L (1999) Efficient gene transfer and long-term expression in neurons using a recombinant adenovirus with a neuron-specific promoter. Gene Ther 6: 1884–1892.

Palmer T.D, Rosman GJ, Osborne WR, Miller AD (1991) Genetically modified skin fibroblasts persist long after transplantation but gradually inactivate introduced genes. Proc Natl Acad Sci USA 88: 1330–1334.

Paulding WR, Czyzyk-Krzeska MF (1999) Regulation of tyrosine hydroxylase mRNA stability by protein-binding, pyrimidine-rich sequence in the 3'-untranslated region. J Biol Chem 274: 2532–2538.

Poewe WH, Wenning GK (1996) The natural history of Parkinson's disease. Neurology 47:S146–152.

Ripps ME. Huntley G.W, Hof PR, Morrison J, Gordon JW (1995) Transgenic mice expressing an altered murine superoxide dismutase gene provide an animal model of amyotrophic lateral sclerosis. Proc Natl Acad Sci USA 92: 689–693.

Rogers JT, Leiter LM, McPhee J, Cahill CM, Zhan SS, Potter H, Nilsson LN (1999) Translation of the alzheimer amyloid precursor protein mRNA is up-regulated by interleukin-1 through 5'-untranslated region sequences. J Biol Chem 274: 6421–6431.

Rosen DR, Siddique T, Patterson D, Figlewicz DA, Sapp P, Hentati A, Donaldson D, Goto J, O'Regan JP, Deng HX, Rahmani Z, Krizus A, McKenna-Yasek D, Cayabyab SM, Berger R, Tanzi RE, Halperin J, Herzfeldt B, Van den Bergh R, Hung W-Y Hung, Bird T, Deng G, Mulder DW, Smyth C, Laing NG, Soriano E, Pericak-Vance MA, Haines J, Rouleau GA, Gusella JS, Horvitz HR, Brown Jr RH (1993). Mutations in Cu/Zn superoxide dismutase gene are associated with familial amyotrophic lateral sclerosis. Nature 362, 59–62.

Sauer H, Rosenblad C, Bjorklund A (1995) Glial cell line-derived neurotrophic factor but not transforming growth factor beta 3 prevents delayed degeneration of nigral dopaminergic neurons following striatal 6-hydroxydopamine lesion. Proc Natl Acad Sci USA 92: 8935–8939.

Tu PH, Raju P, Robinson KA, Gurney ME, Trojanowski JQ, Lee VM (1996) Transgenic mice carrying a human mutant superoxide dismutase transgene develop neuronal cytoskeletal pathology resembling human amyotrophic lateral sclerosis lesions. Proc Natl Acad Sci USA 93,: 3155–3160.

Warita H, Itoyama, Y, Abe K (1999) Selective impairment of fast anterograde axonal transport in the peripheral nerves of asymptomatic transgenic mice with a G93A mutant SOD1 gene. Brain Res 819:120–131.

Wickham TJ, Tzeng E, Shears LL 2nd, Roelvink PW, Li Y, Lee GM, Brough DE, Lizonova A, Kovesdi I (1997) Increased in vitro and in vivo gene transfer by adenovirus vectors containing chimeric fiber proteins. J Virol 71: 8221–8229.

Williamson TL, Cleveland DW (1999) Slowing of axonal transport is a very early event in the toxicity of ALS-linked SOD1 mutants to motor neurons. Nature Neurosci 2: 50–56.

Wong PC, Pardo CA, Borchelt DR, Lee MK, Copeland NG, Jenkins NA, Sisodia, SS, Cleveland DW, Price DL (1995). An adverse property of a familial ALS-linked SOD1 mutation causes motor neuron disease characterized by vacuolar degeneration of mitochondria. Neuron 14: 1105–1116.

Wong, PC, Borchelt, D.R, Lee M.K, Pardo CA, Thinakaran G, Martin LJ, Sisodia S S, Price DL (1998) Familial amyotrophic lateral sclerosis and Alzheimer's disease. Transgenic models. Adv Exp Med Biol 446,:145–159.

Wood MJ, Charlton HM, Wood K J, Kajiwara K, Byrnes AP (1996) Immune responses to adenovirus vectors in the nervous system. Trends Neurosci 19:497–501.

Yang Y, Nunes F.A, Berencsi K, Furth EE, Gonczol E, Wilson JM (1994) Cellular immunity to viral antigens limits E1-deleted adenoviruses for gene therapy. Proc Natl Acad Sci USA 91: 4407–4411.

Zennou V, Petit C, Guetard D, Nerhbass U, Montagnier L, Charneau,P (2000). HIV-1 genome nuclear import is mediated by a central DNA flap. Cell 101: 173–185.

Zennou V, Serguera C, Sarkis, C, Colin P, Perret E, Mallet J, Charneau P (2001) The HIV-1 DNA flap stimulates HIV vector-mediated cell transduction in the brain. Nature Biotechnol 19,:446–450.

Zhang B, Tu P, Abtahian F, Trojanowski JQ, Lee VM (1997) Neurofilaments and orthograde transport are reduced in ventral root axons of transgenic mice that express human SOD1 with a G93A mutation. J Cell Biol 139:1307–1315.

# Subject Index

Printing (Computer to Plate): Saladruck Berlin
Binding: Stürtz AG, Würzburg